LED Circuits & Projects

by

Forrest M. Mims, III

Howard W. Sams & Co., Inc.
4300 WEST 62ND ST. INDIANAPOLIS, INDIANA 46268 USA

FIRST EDITION
FIFTH PRINTING—1979

International Standard Book Number: 0-672-21006-1
Library of Congress Catalog Card Number: 73-83371

Printed in the United States of America.

Preface

In the process of writing my previous book, *Light Emitting Diodes* (© Howard W. Sams & Co., Inc., 1973), it soon became apparent that there was too much material to include in one book. The finished book was much longer than anticipated, so it was then limited to theory, basic circuitry, and operation. This second book then covers the application for light emitting diodes (LEDs).

Like the previous publication on LEDs, this book could not have been written without the generous assistance of many individuals and firms. I am particularly indebted to John W. Hall, II, of the General Electric Company for numerous sample LEDs, photographs, and encouragement. Dick Lundgren, also of GE, secured permission to reprint information contained in the company's *Solid State Lamp Manual.*

One of the most difficult areas in the field of optoelectronics deals with the problem of measuring LED power output, and several firms came forward with invaluable assistance in this area. John Hall, II, of GE and Dr. Richard Glicksman of RCA both provided calibrated LEDs. Mr. Dave Goedel of Optronic Laboratories generously consented to evaluate LEDs for this and subsequent books. Mr. Neil S. Bernstein of EG&G provided a calibrated, flat-spectral response detector. Metrologic Instruments, Inc. provided one of its economical photometers. Pete Keller of Tektronix, Inc. loaned an accurately calibrated J16 digital radiometer from his firm. The assistance provided by these men and their firms has made an important contribution to the preparation of this book and succeeding books, now in preparation, on the subject of optoelectronics.

H. Edward Roberts, president of Micro Instrumentation and Telemetry Systems, Inc., permitted me to photograph LED displays used by his company in the manufacture of digital calculators and clocks. Mr.

Roberts also granted permission to reproduce the circuit diagrams for the Opticom LED communicator described in Chapter 5.

Monsanto sales representative Dwayne Frye provided invaluable assistance in obtaining samples of new LEDs and literature. Jack Pintar of Litronix provided samples of LEDs from his firm. Jim Graham supplied samples of Hewlett-Packard LEDs, technical literature, and photos. Ken Horton, Bob Cole, Lin Wetterau, and Ted Meador of Texas Instruments provided technical advice, literature, photographs, and sample LEDs.

James P. Miller, chairman of the Southwest Research Association, provided valuable assistance with the optical radar range equations in Chapter 6.

Finally, I continue to be grateful for the patient and diligent cooperation of my wife Minnie. Besides her invaluable assistance in proofreading and typing the drafts and final manuscript, she has provided the understanding and love which is so essential to the completion of a book. To these many individuals and companies, especially to my wife and partner, I offer my sincere gratitude.

FORREST M. MIMS, III

Contents

CHAPTER 4

CHAPTER 5

CHAPTER 6

CHAPTER 7

CHAPTER 1

Light Emitting Diodes

In recent years, a family of semiconductor devices capable of emitting visible and infrared light has become available. Semiconductor light emission was first discovered in 1907 by H. J. Round, but it was not until 1960 that efficient light generation in a semiconductor was obtained.

The material which made possible this high-efficiency light generation is gallium arsenide (GaAs). Like germanium, silicon, and certain other materials, GaAs is a semiconductor and possesses electron conduction properties intermediate between those of conductors and insulators. But GaAs is unique in that it possesses a very high efficiency for converting an electrical current directly into light.

Actually, all semiconductor devices possess some ability to convert a current flow into visible or infrared light. A junction of p and n material is not always necessary, but the most efficient light generation happens to occur at such an interface. An ordinary silicon or germanium diode will produce a tiny amount of visible infrared light when forward biased. One diode which is able to produce such light somewhat more efficiently than others is the 1N3070. This silicon diode will be seen to emit a distinct glow when biased at 100 mA forward current and observed through an infrared image converter (Fig. 1-1). As might be expected, some 1N3070 units perform better in this application than others.

For most practical purposes, silicon and germanium are not suited for infrared generation. The radiation they generate is very inefficient and accompanied by a considerable amount of heat. However, GaAs generates light with very high efficiency. The reason for this efficiency can be understood with an explanation of semiconductor light emission.

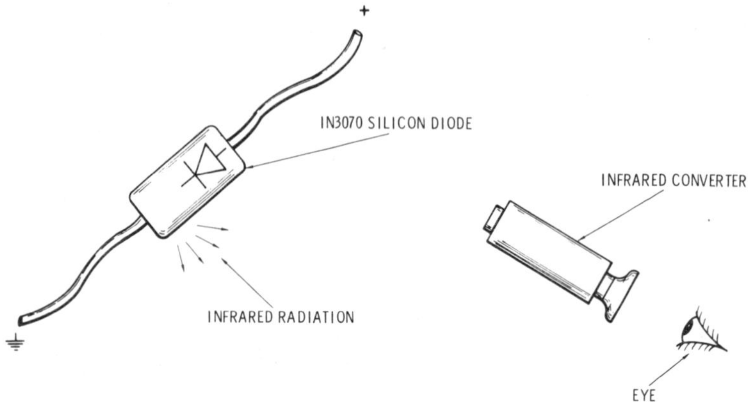

Fig. 1-1. Recombination radiation from common silicon diode.

SEMICONDUCTOR LIGHT EMISSION

The smallest constituent of an element is the atom. As shown by the diagram of a neon atom in Fig. 1-2, atoms contain a variety of subatomic particles. For our purposes the most important of these is the electron. While Fig. 1-2 shows the electrons as finite particles, it is more correct to think of an electron as a negatively charged cloud encircling the nucleus.

Fig. 1-2 shows the electrons of the neon atom occupying several discrete levels. All atoms follow this guideline, and those with more electrons than the neon atom have more levels. The outermost level is called the *valence* band, and the number of electrons it contains determines the stability of the atom. In all atoms, except hydrogen, the valence band must have a full complement of eight electrons to have the highest possible degree of stability. Atoms with a filled valence band are so stable they are called *inert*.

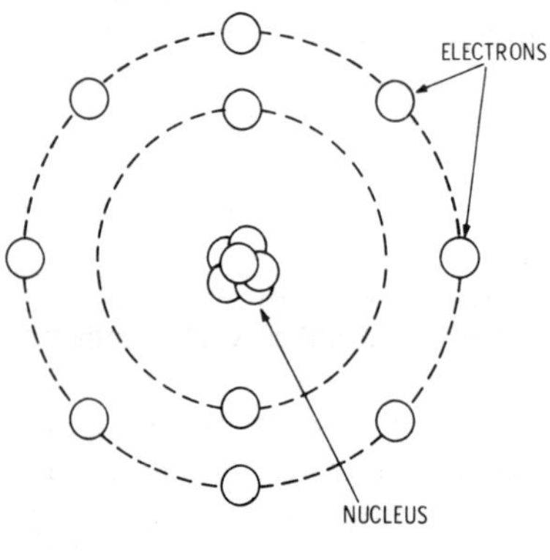

Fig. 1-2. Neon atom.

Only the so-called noble gases (helium, neon, argon, krypton, xenon, and radon) are inert, and all other elements have vacancies in their valence band. The ability of an element to conduct electricity is directly related to the number of electrons in the valence band. Elements with from one to three electrons in the valence band are chemically active and good electrical conductors, but elements with from five to eight electrons in the valence band are inactive and poor conductors.

Now that the basic structure of the atom has been explained, we can go on to describe how an atom emits light. An atom can absorb external energy such as heat, light, X-rays, or gamma rays. The atom compensates for this energy contribution by causing one of its valence electrons to move to a higher energy level. This new level is called the *conduction* band, and it is normally unoccupied. Under normal circumstances, the electron cannot occupy a site between the valence and conduction bands. For this reason the space is called the *forbidden gap.*

The stimulated atom may retain the electron in the conduction band for a time ranging from nanoseconds to days. Sooner or later, however, the atom resumes its normal state when the stimulated electron falls back to the valence band (sometimes referred to as the *ground state*). When this occurs, equilibrium is preserved as the stimulated electron gives off the energy it absorbed in the form of light or heat. This radiated energy is called *recombination radiation* since an electron has combined with the hole it once occupied and, in the process, it has emitted radiation..

In GaAs recombination radiation is almost exclusively near infrared. Fig. 1-3 shows the relationship of near infrared to the visible light spectrum and gives several common LED wavelengths. Since little heat is produced, GaAs is a very efficient source of light. In fact, in properly prepared GaAs, almost every stimulated electron produces a photon of near infrared. The ratio of emitted photons and incoming electrons is called the *quantum* efficiency. If, for example, 8 of every 10 incoming electrons stimulates the emission of a photon, the semiconductor has a quantum efficiency of 80%.

While a material like GaAs may have a quantum efficiency of nearly 100%, much of the light generated within the crystal never emerges from the surface. Two major loss mechanisms are absorption within the crystal and internal reflection at its surface. Blockage caused by electrical contacts and the diode mounting package contributes significant loss as well. For this reason, quantum efficiency is usually expressed in terms of internal and external efficiency, with the latter always being less.

Power conversion efficiency is another important LED output parameter. Power efficiency is a figure of merit for the ratio of electrical

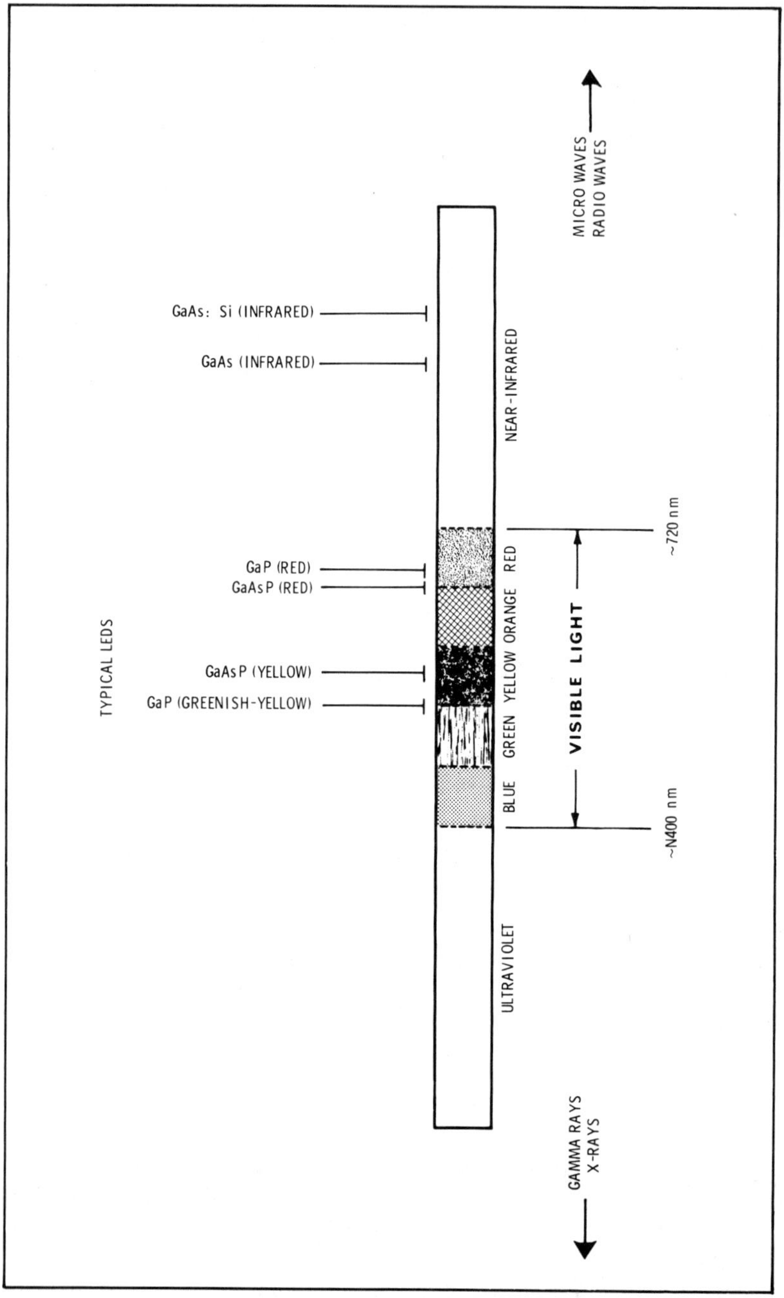

Fig. 1-3. Electromagnetic spectrum.

input to optical output power. Typical GaAs LEDs have a power efficiency of several percent and the use of an integral reflector and epoxy encapsulant increases this value to more than 5%. Expensive hemispherical LEDs have power efficiencies of 10%.

The efficiency values so far discussed are for operation at room temperature. Cooling increases the transmission and the external efficiency of GaAs and other light emitters. A hemispherical diode, for example, may have a power efficiency of more than 25% at the temperature of liquid nitrogen (−196 °C). Cooling also reduces the wavelength of an LED by about 0.25 nm per degree Celsius (formerly centigrade).

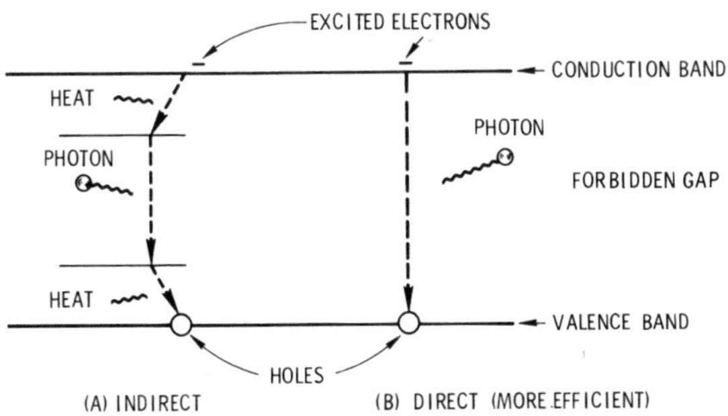

Fig. 1-4. Radiative recombination.

Fig. 1-4 shows the process of radiative recombination in an efficient light generator, such as GaAs, and an inefficient light source, such as silicon or germanium. In (A), the transition of an electron from the excited conduction band to the valence band is in a series of steps. An optical photon may be emitted at only one of these steps, while heat is given off at the others. In (B), the electron transition is direct, and an optical photon is emitted without heat. While light emitting diodes are made from both indirect and direct bandgap semiconductors, the latter are preferred for higher efficiency. The wavelength of the light emitted during radiative recombination can be easily calculated; it is directly related to the energy separation between the valence and conduction band in electron volts. The formula is:

$$\lambda = \frac{hc}{E} \qquad \text{(Eq. 1-1)}$$

where,

h is Planck's constant (6.63×10^{-34} joule-seconds),
c is the velocity of light (3×10^{14} micrometers per second),
E is the energy in joules that separates the valence and conduction bands.

The equation can be simplified by converting to electron volts:

$$\lambda = \frac{1237 \text{ nanometers}}{E_g \text{ (electron volts)}} \qquad \text{(Eq. 1-2)}$$

With a bandgap separation of about 1.34 electron volts, GaAs should emit radiation with a wavelength of 923 nanometers, according to Equation 1-2. This is close to the actual wavelength of about 905 nanometers.

Ideally, the radiation emitted during recombination would be of a single, monochromatic wavelength; if the bandgap between the valence and conduction bands was a discrete distance, this would be the case. The bandgap cannot be this precise, however, since the electrons encircle the atom nucleus as an electrostatic field. While there is a probability that the bandgap in GaAs, for example, will be 1.34 electron volts, some atoms of GaAs may have a bandgap of 1.32 or 1.36 electron volts. More precisely, the bandgap of an individual atom may vary slightly with time due to the fuzzy boundaries of the electrostatic field.

Because of the bandgap uncertainty, the wavelength emitted by GaAs is some 25 to 40 nanometers wide. As we shall see in Chapter 7, this figure can be reduced considerably by means of laser action.

LIGHT EMITTING DIODES

An efficient light emitter such as GaAs can be stimulated to emit light by an electron beam, a bright light source, or electron injection. It happens that the direct injection of electrons is the simplest method. Electron injection is facilitated by fabricating the GaAs into a pn junction diode. When the diode is connected to a source of current, electrons are injected into the n region. To cross the potential barrier formed by the junction, the injected electrons are transferred to the conduction band, as shown in Fig. 1-4. When the junction is crossed, the electrons drop back to holes in the valence band and, in the process, give off their excess energy as recombination radiation.

Radiative recombination in a GaAs junction diode is summarized in Fig. 1-5 by means of a chart called an *energy-level diagram.* The valence and conduction bands on each side of the pn junction are con-

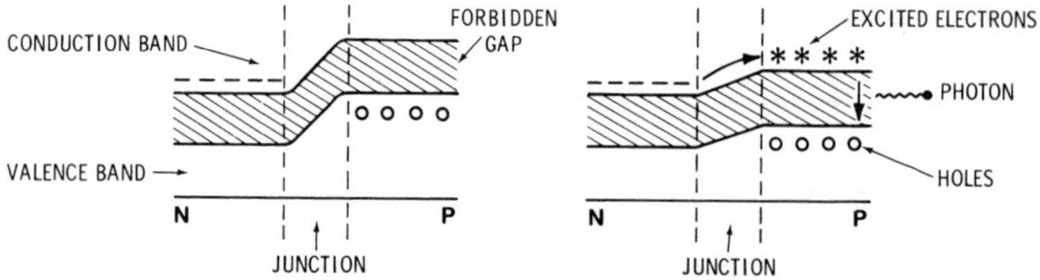

Fig. 1-5. Energy-level diagram of a pn junction.

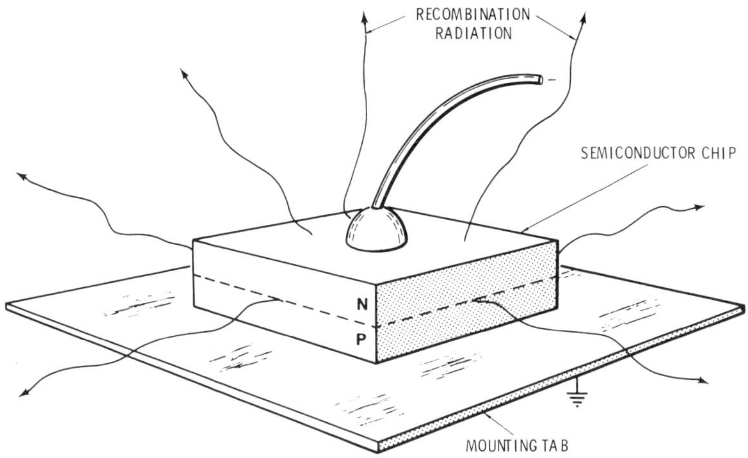

Fig. 1-6. Basic LED structure.

nected by sloping lines representing the junction barrier to electron flow. When electrons are injected into the n side by a current source, the potential barrier of the junction is reduced, and current flow occurs.

A basic planar, or flat, LED is shown diagrammatically in Fig. 1-6. Numerous variations in construction, physical appearance, and electrode configuration are possible. A typical diode consists of p-type semiconductor, with a diffused layer of n-type material. The p-type layer is generally bonded to a metal header, since it tends to be less transparent to the diode radiation and more radiation can escape from the chip with the n region on top. An electrode wire is bonded to the n-type layer to complete the necessary electrical connections. In operation, recombination radiation generated at the pn junction is emitted from all portions of the diode not blocked by electrodes. Frequently, the metal header is highly reflective; so the radiation passing through the p region can be reflected back toward the diode surface.

The electrical characteristics of an LED are similar to those of other junction diodes. As shown in Fig. 1-7, current rises steeply with an applied voltage. If the forward current-voltage curve begins to taper at higher current values, overheating is occurring, and the diode should be cooled or the current reduced to prevent degradation of output power or possible device destruction.

LED CHARACTERISTICS

The LED possesses the characteristic voltage-current relationship of most junction diodes, i.e., current increases rapidly with voltage remains nearly constant. Fig. 1-7 shows a graph of voltage versus current for a typical GaAs LED. There is a limit to forward current and reverse voltage an LED can withstand without permanent damage. Most commercial LEDs can be forward biased at up to 50 or 100 mA without heat sinking, but higher current levels usually require a heat sink to prevent thermal damage. Reverse voltages of a few volts can be tolerated in most cases.

The light output of a typical LED is linear with applied current. Fig. 1-8 is a graph showing the output of a red-emitting, gallium arsenide phosphide (GaAsP) LED. The graph was made by pointing the LED into a calibrated Tektronix J16 Photometer and applying current to the LED with a calibrated power supply having a current-indicating accuracy of 0.1%. Notice that the light output of the diode is perfectly linear until a current of 65 mA is reached. Beyond 80 mA, the light output falls well below the peak value.

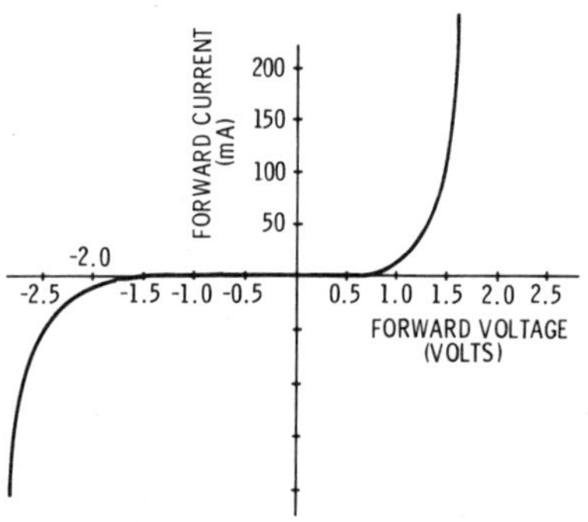

Fig. 1-7. Graph of voltage and current in an LED.

The drop in light output is caused by overheating of the LED chip. The diode used in the test, an epoxy encapsulated unit, would have operated at higher current levels if it had been installed on a metal header connected to a heat sink. But this particular LED is intended for low-current operation as an indicator lamp, and it fills this role nicely. In fact, this diode is rated for a peak continuous forward current of 50 mA. As Fig. 1-8 shows, it still operates in an approximately linear region at 80 mA.

The linear relationship of input current to light output in an LED is useful in many applications. Several such applications, including amplitude-modulated voice communications and optical potentiometers, will be described in detail in subsequent chapters.

Another important characteristic of LEDs is their very rapid response time. Earlier, we noted that a stimulated electron can remain in an excited state in the conduction band for a time ranging from nanoseconds to a few days. In GaAs and GaAsP, the lifetime of a stimulated electron can be less than a single nanosecond. For this reason, a GaAs or GaAsP LED can begin emitting recombination radiation within a nanosecond of an applied current pulse. Depending on the structure of the LED, this time may vary from well under a nanosecond to 10 or more nanoseconds. LEDs compensated with a silicon dopant are much slower than GaAs and GaAsP, but they have much higher efficiencies. Nevertheless, even their response time of about 300 nanoseconds is faster than many other light sources.

The very fast response time of the LED makes it an ideal source for high-frequency optical communications. A typical GaAs LED, for example, can easily be modulated at a rate of 100 MHz. Gallium arsenide LEDs, compensated with silicon for higher conversion efficiency, can be modulated at a 1-MHz rate.

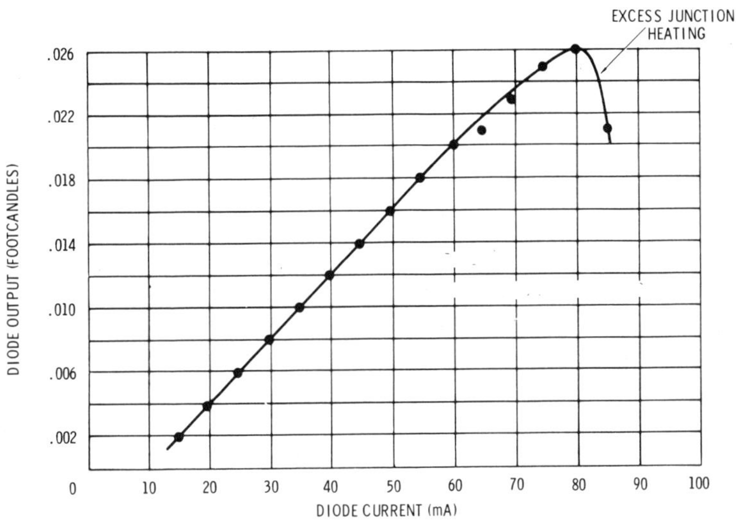

Fig. 1-8. GaAsP light output vs. forward current.

Another important advantage of LEDs is a virtually unlimited operating life. Incandescent lamps do well to operate beyond 10 thousand hours, but inexpensive LEDs are rated for lifetimes of 100 thousand hours and more. Several accelerated life studies have predicted typical LEDs will operate 100 years before their light output drops to half its initial value! Even when providing lessened light output, an LED is still valuable as, for example, an indicator lamp. An incandescent lamp ceases operation in an instant, but an LED continues to produce usable light with only a very gradual degradation over a period of many years.

The operating life of an LED is directly related to its duty cycle (total "on" time) and junction current density. For this reason, diodes operated at high-current levels to achieve greater power outputs will exhibit reduced operating life. Nevertheless, their life is far longer than most equivalent sources.

Still another important characteristic of the LED is its relatively narrow spectral bandwidth. This permits semiconductor indicator lamps to be made which emit yellow, amber, green, and red without the intervening filter which is necessary for incandescent lamps to emit discrete colors. A narrow spectral output is important in communications applications for two reasons. First, LEDs can be made with a peak output corresponding to the peak sensitivity of readily available detectors. This means more of the light being transmitted has a chance of being detected than in systems employing a wide-spectrum light source. Second, narrow spectral width means a filter which transmits only the LED wavelengths can be placed over the detector to help eliminate interference from outside light sources.

Finally, LEDs possess the inherent physical advantages of most solid-state devices. They are relatively inexpensive and easy to make. They do not require a very high operating voltage, they have a very high quantum efficiency, and they will operate in a wide range of temperatures. Ordinary commercial LEDs, for example, can frequently be operated at temperatures ranging from that of liquid nitrogen (−196 °C) to boiling water (100 °C). Since the operating characteristics of LEDs change with temperature, it is necessary to adjust the current level to prevent damage. As temperature is decreased, efficiency increases and wavelength decreases, at a rate of about 0.25 nanometers per degree Celsius.

LEDs are also very sturdy. This last characteristic was vividly demonstrated to the author during an experiment with a miniature LED telemetry system installed in a small rocket. The LED transmitted a modulated signal to a sensitive ground receiver in an experiment to determine the potential of an air-to-ground infrared telemetry link. After one of the launches, the parachute on the rocket failed, and the instrument section was entirely destroyed. The LED was the only component that remained intact.

VISIBLE LEDS

Millions of visible LEDs are used annually as indicator lamps and readout displays. Fig. 1-9 shows a special test board which uses a bank of LEDs to indicate the operating status of circuits in an electronic terminal. Until recently, almost all visible LEDs were GaAsP red emitters, but recent advances in semiconductor technology have permitted the production of relatively efficient, and economical, green and yellow diodes.

Gallium arsenide phosphide (GaAsP) red emitters are by far the most inexpensive LEDs available, and one manufacturer markets machine-manufactured units for only 10¢ each, when purchased in quantities of a million. Their peak wavelength of emission is about 660 nm and they have the very fast response times of GaAs units.

Gallium phosphide (GaP) red emitters are more efficient than GaAsP units, but the eye is less sensitive to their 690-nm wavelength. Also, GaP emitters are difficult to manufacture due to the problems in producing consistently good quality GaP.

Fig. 1-9. LEDs used as indicators in electronic equipment test board.

GaP can also be used for making green-emitting LEDs. These diodes emit a range of wavelengths very near the peak sensitivity region of the human eye (555 nm) but have a low power conversion efficiency. Yellow emitters are made from GaAsP. Again, their power efficiency is low, but their visual efficiency is high.

Green and yellow emitters are still more expensive than equivalent red emitters. Prices continue to drop as production techniques improve, and it is now possible to buy a green emitter for about $1.19 from some electronics discount suppliers. In 1969 this identical diode sold for $125.00! A British LED manufacturer is now selling green-emitting GaP LED chips at only 15¢ a die in 100-thousand quantities, a rate only 20% below red-emitting GaAsP chips. Nevertheless, fundamental differences between GaP and GaAsP make it highly unlikely than green and yellow diodes will ever be as inexpensive as red emitters. Green and yellow emitters will always be in demand, however, since there are many applications where a color other than red is highly desirable.

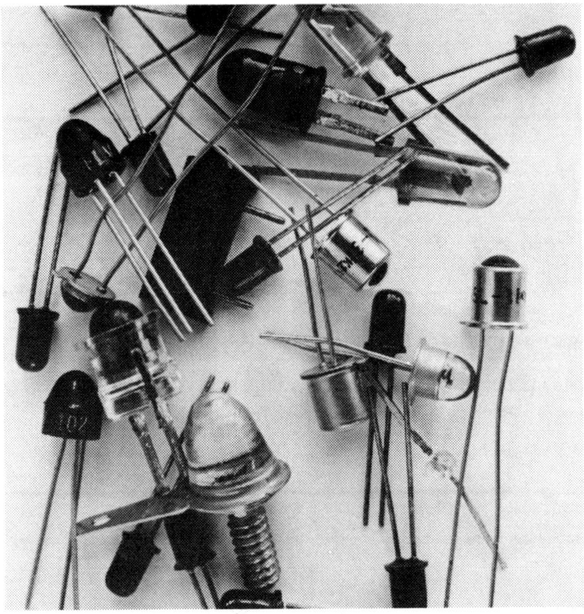

Fig. 1-10. Assortment of visible LEDs.

As shown in Fig. 1-10, visible LEDs are available in a wide variety of configurations. Inexpensive indicator units are invariably epoxy encapsulated, while high-power units incorporate both epoxy encapsulation and a metal header. A variety of lens colorations are available. Some diodes have a clear encapsulant while others use a dye with the same color as the LED wavelength. The transparent encapsulant provides a very bright point-source of light, but viewing is uncomfortable in some situations. Also, in the presence of bright ambient light, reflections from the LED chip and its connection wires tend to reduce the apparent brightness of the diode. These problems are overcome by adding a diffusing dye to the encapsulant. Rather than a bright

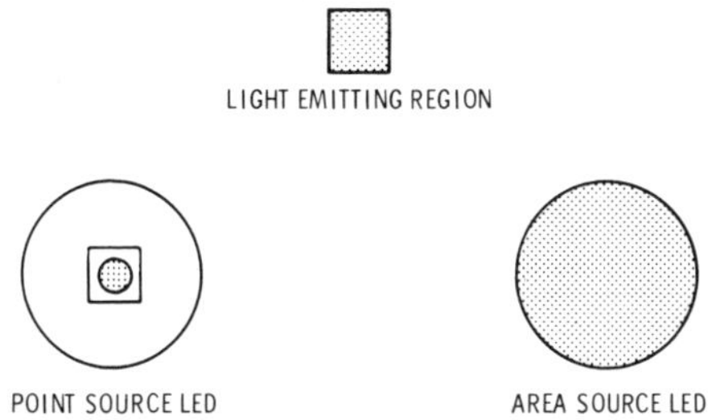

Fig. 1-11. Comparison of point and area source LEDs.

point source, the entire encapsulant glows with the color of the LED. A point-source and area-source LED are compared in Fig. 1-11.

INFRARED LEDS

Infrared emitting LEDs have been made from a variety of semiconductors, but the most popular are GaAs and GaAs:Si. These diodes emit at wavelengths centered at about 905 and 940 nanometers respectively. Since these wavelengths are normally considered beyond the range of visible light, it can be argued that infrared-emitting diodes should not be referred to as *light emitting diodes.* Indeed, sometimes infrared emitters are labeled IREDs (infrared-emitting diodes). For convenience, this book places both visible- and infrared-emitting diodes under the general category of LEDs. For all practical purposes, the "light" emitted by infrared LEDs is invisible to the human eye, but its properties are very similar to that emitted by visible LEDs. It should be noted that even a 905-nanometer beam can be seen if it is sufficiently intense. The peak wavelength sensitivity of the eye is usually given as about 750 nanometers, but the radiation from GaAs lasers at 905 nanometers is clearly visible as a distinct red when viewed in subdued ambient light.

The GaAs and GaAs:Si LEDs each have relative advantages and disadvantages. While GaAs units have very fast response times and can be modulated at 100 MHz, they are less efficient than GaAs:Si types. The lack of efficiency is partially due to the absorption of recombination radiation within the semiconductor making up the diode. Amphorteric doping of both sides of the junction changes the wavelength of the recombination radiation to about 940 nm, which corresponds to a bandgap energy *below* that of the highly absorbing GaAs without silicon. The result is that more of the radiation generated at the junction can pass through the semiconductor without being absorbed.

The GaAs:Si LEDs can have power conversion efficiencies of 10% and more, but their response time is considerably slower than GaAs units. While GaAs diodes may exhibit rise times of a nanosecond or less, silicon-compensated units have rise times of 300 ns or more.

Summing up, for applications requiring a high-modulation rate, GaAs LEDs are preferred. But, for high-peak power and efficiency, the silicon-compensated units are best.

Infrared LEDs are available in less configurations than visible units, but there is sufficient variety for almost all applications. Fig. 1-12 shows an assortment of typical infrared emitters. Most infrared units incorporate a metal header for heat sinking and an epoxy encapsulant to reduce the critical interface problem. Some diodes incorporate vari-

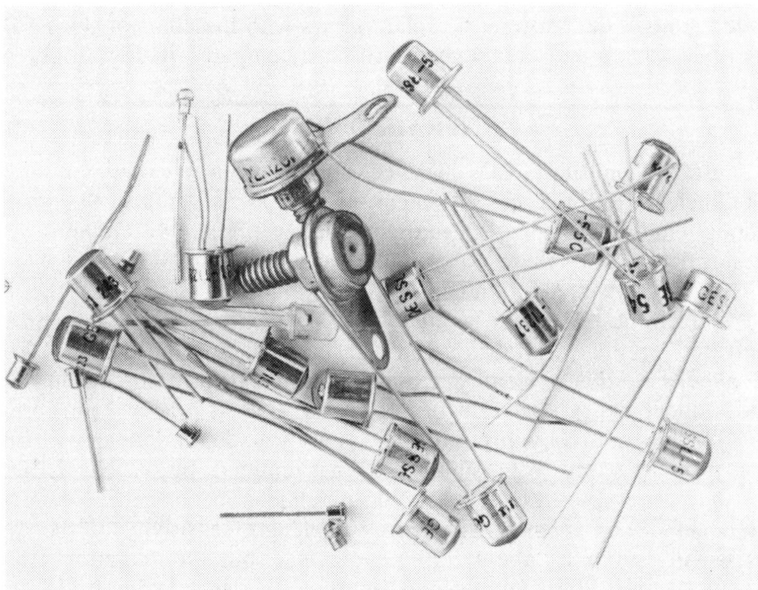

Fig. 1-12. Assortment of infrared LEDs.

ous combinations of internal reflectors and lenses to collect more of the radiation of the diode into a narrow beam. These diodes are valuable in applications requiring an intense beam of near infrared.

LED CONFIGURATIONS

LEDs are available in literally dozens of configurations, but most utilize some form of *immersed optics* for better light extraction efficiency. Gallium arsenide and other light-emitting semiconductor compounds are characterized by a very high index of refraction. For example, GaAs has a refractive index higher than that of diamond. The interface between two substances with different refractive indexes is highly reflective to oncoming radiation arriving at an angle greater than the *critical angle.* In GaAs, this angle is about 16 degrees; therefore, only light from within the crystal, which strikes the GaAs-air interface within a 32-degree cone, is emitted from the crystal. The rest is reflected back into the semiconductor and lost.

Some very expensive LEDs have a light-emitting chip which is formed into a hemisphere. Since light always strikes the crystal-air interface at a near normal angle (well within the required 32-degree cone), there is no surface reflection loss. Even though there is some loss due to absorption within the relatively thick crystal dome, these

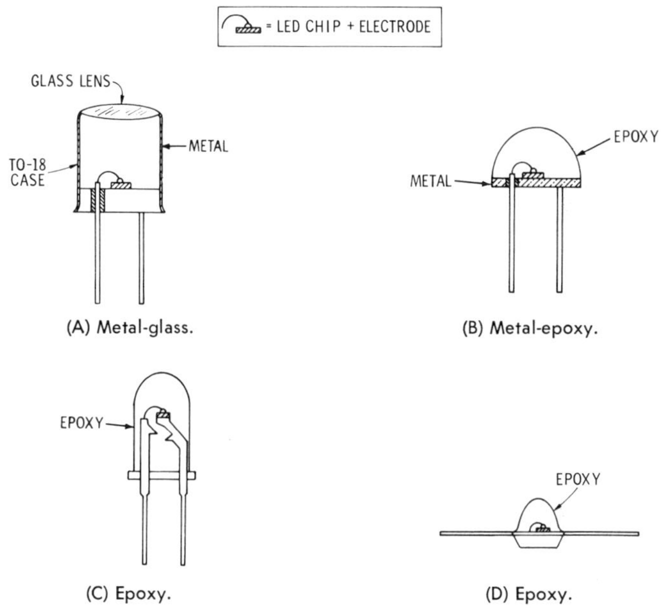

Fig. 1-13. Common LED configurations.

hemispherical diodes are several times more efficient than conventional flat emitters.

An inexpensive way to reduce the critical interface problem, without using a domed diode structure, is to encapsulate the crystal in a clear material having an index of refraction between that of the semiconductor and air. For several years now, practically all visible LEDs have been made using this technique. Clear epoxy increases the critical angle to perhaps 30 degrees and provides about 2½ times more light output than an unencapsulated diode.

The most common LED configurations are shown in Fig. 1-13. There are several variations on each of the various configurations, and external optics can be employed with an LED.

In addition to improving the light extraction from an LED chip, an epoxy encapsulant can be geometrically shaped to produce a variety of output beams. A rounded surface close to the chip will give a broad beam, while a small rounded surface farther away from the chip will give a narrow beam. To improve viewing ability, the encapsulant is frequently dyed, diffused, or both. This causes the light from the chip to spread throughout the epoxy and converts the LED from a tiny point source to a large area source. Point sources are useful when it is necessary to employ an external lens, but area sources are far more comfortable to observe. Because of the subjective nature of human vision, an area source producing somewhat less light than a point source is generally more noticeable and easier to see.

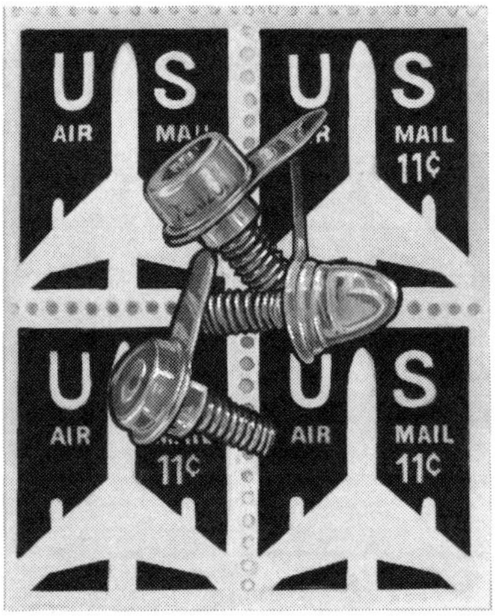

Fig. 1-14. High-power LEDs with stud mounts.

Since visible LEDs are commonly used as indicators, economy is essential and hermetic sealing is usually not provided. For military applications, however, ceramic and glass, or metal and glass packages are provided.

Infrared emitters are available in both hermetically sealed metal cans with a glass lens or window and various encapsulated forms. Metal-epoxy units are becoming more popular due to their low cost and the improvement in extraction efficiency. Even metal-glass packaged diodes frequently employ a small droplet of clear epoxy directly over the chip itself to enhance extraction efficiency.

High-power infrared emitters and some high-power visible diodes are mounted in a high thermal efficiency package employing a large threaded stud for a heat sink (Fig. 1-14). These diodes can be operated at continuous currents of 200 mA or more without external heat sinking. By mounting the stud to a heat sink, even higher current levels are possible.

CHOOSING VISIBLE LEDS

With dozens of different light emitting diodes available, choosing an LED for a particular application can be time consuming. One major manufacturer issues a quarterly catalog describing its LED products in detail, and the spring 1973 edition described more than 100 LED products in 168 pages of diagrams, charts, and specification listings. The problem is compounded by a lack of industry standardization. Several companies "second source" one another on certain fast-selling LEDs and LED displays, but most LEDs are unique components made by only one manufacturer. A company planning to employ a particular LED in a product must use care in selecting a device with critical operating parameters, since the manufacturer may decide to cease production.

Most LED users would like to see LED manufacturers standardize their products much like transistors, diodes, and other semiconductors. In this way, a customer could purchase a particular device by the same type number from several manufacturers. So far, the manufacturers have been reluctant to standardize their products due to the highly competitive nature of the market. Hope is in sight, however, and some observers compare the present LED market with the early transistor period and predict eventual industry standardization of most popular LEDs.

Visible LEDs are generally much easier to choose than infrared emitters. In display and indicator applications, the key rule is *look before buying*. Most manufacturers supply sales representatives with a display board containing a variety of LEDs. The diodes can be operated by a master switch and viewed in actual operating conditions in any desired light level. An alternate approach is to obtain several sample devices from the manufacturer and try them out before making a final selection.

Experimenters and other small-quantity LED users need not be so careful. A dozen or more electronics mail-order suppliers offer numerous types of LEDs and LED displays at bargain-basement prices. Many of these dealers advertise in the pages of electronics experimenter magazines.

Most applications for visible LEDs employ red-emitting GaAsP diodes, but, as prices continue to come down, green and yellow units will become more popular. Also, improved gallium phosphide semiconductor technology is making efficient GaP red emitters cost competitive with the older GaAsP units. The system designer must understand the relative operating and output characteristics of these various visible emitters before employing them in a circuit. For example, a green GaP LED cannot be directly substituted for a red GaAsP unit, since the red device requires a forward voltage of only about 1.8 volts, while the green unit requires about 3.5 volts. Unless the voltage is adjusted upward, a direct substitution will result in a barely visible glow from the green diode.

Color or hue of the output radiation is another important consideration. The eye is primarily sensitive to wavelengths ranging from about 400 to 720 nm. Very bright sources beyond these wavelengths can be seen also, but viewing may be hazardous to the retina due to the high-power level required. The peak sensitivity of the eye is in the green at about 555 nm. A red light at 650 nm, with the same power as a green light at 555 nm, will appear only about one-tenth as bright to the eye. This is due to the decrease in visual response at wavelengths below and above the green portion of the spectrum.

Fig. 1-15 shows the relative response of the typical human eye (photopic or color vision), and the typical spectral output of the most common LEDs. This curve is called the *photopic luminosity curve* and most individuals have a similar visual response. About one of every fifty males experiences a reduced sensitivity to GaAsP and GaP red emitters.

From the curve in Fig. 1-15, it would appear that green and yellow emitters offer significantly higher efficiencies than red sources. From a visual standpoint, this is certainly true, but presently available green and yellow diodes have only a fraction of the quantum efficiency of red units. For example, a good GaP green emitter may have an external quantum efficiency of only 0.1%, while typical GaAsP and GaP red LEDs have external quantum efficiencies of 0.5% and 3% respectively. Therefore, the very high visual efficiency of green diodes is offset by the very high quantum efficiency of red diodes.

Still another consideration is the subjective nature of a visible diode hue or color. To some observers, for example, an amber or orange

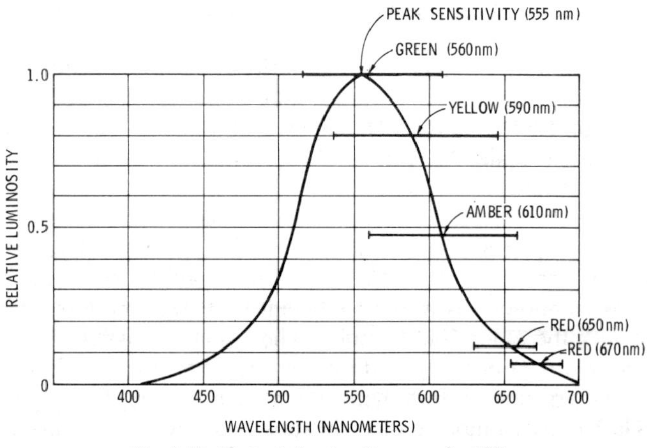

Fig. 1-15. Photopic luminosity curve for LEDs.

LED appears red. Green diodes are more difficult to evaluate visually, since most commercial units appear greenish yellow. Variations in the ambient light level or the presence of a nearby yellow emitter can cause a green diode to appear more green. Often, however, the color of a "green" emitter is more greenish yellow than green. Phosphor-coated GaAs *up converter* LEDs produce a very pure green of a slightly lower wavelength than GaP diodes but have lower efficiency.

The difficulty in accurately defining the hue of green, and some other visible LEDs, stems from the sloppy bandgap of the indirect semiconductors used to make the diodes in question. The poorly defined bandgap and emission at different bandgaps on either side of the junction give rise to a wide emission spectrum centered at the primary wavelength of interest. A good way to verify the wavelength range of a typical visible emitter is to observe the unit through an inexpensive clear plastic diffraction grating in a darkened room. In the case of a GaP green emitter, a range of colors extending from the brilliant green to the yellowish orange will be seen. The colors combine to give the greenish yellow (or yellowish green) seen without the grating. Inexpensive gratings, as well as other optical equipment suitable for use with LEDs, are available from Edmund Scientific Company (300 Edscorp Bldg., Barrington, NJ 08007).

CHOOSING INFRARED LEDS

Choosing an infrared LED can be far more difficult than selecting visible emitters. The most obvious difficulty is the invisible nature of the infrared radiation of the unit. At present, the power ratings given for infrard LEDs can rarely be relied upon. For example, the G.E. SSL-55C emits approximately 6 mW at 100 mA forward bias. Similar diodes, manufactured by two other LED companies and rated for identical output values, actually produce significantly less power. As a result, a system employing the SSL-55C, a high-quality LED, cannot necessarily be operated with second-source diodes, supplied by competitors, unless the reduction in output power is permissable.

A spokesman at one of the largest optical calibration facilities in the world notes that the major difference in specified and actual output for many infrared LEDs is not necessarily an intentional move by the manufacturers. He points out that the great difficulty in making accurate measurements of optical power is due to differences in the emission pattern of various LEDs and utilization of different types of calibrated detectors and measurement techniques.

Another problem with infrared emitters is beam homogenity. Infrared units are almost always used in applications requiring a detector, and power density at the detector should be high for best results. Therefore, one of two diodes, rated for the same optical output, may be unsatisfactory in a particular application if one diode employs an integral lens for a narrow beam, and the other a small reflector for a wide beam. The problem becomes more complex when one considers that infrared LEDs with a self-contained lens emit most of their radiation into an off-axis halo surrounding a primary beam.

The problem of beam homogenity in infrared LEDs is discussed in more detail later in this chapter. The important point of this discussion is that a far better appraisal of the characteristics of an infrared emitter can be had from actual operating tests, rather than data sheet surveys. Of course manufacturers' data sheets supply important information about LED characteristics, but as in the case of visible LEDs it is wise to try a diode before purchasing it in quantity.

Many applications for infrared emitters are possible, and several, including optical communications and detection systems, are described in detail in subsequent chapters. For applications requiring very high-power output, semiconductor diode laser sources should be considered. These devices, which are actually modified LEDs, are described in detail in Chapter 7.

MOUNTING LEDS

In many applications LEDs can be mounted by their connection leads, with plastic mounting grommets, or with clips. When an external lens is required, however, special mounting techniques may be necessary. One good way to pair an LED with an external lens is to install both components in a hollow cylinder. The lens can be epoxied to a short tube which slides inside the main cylinder. This permits the lens to be conveniently focused without moving the LED. An alternate approach is to secure the lens between two retainers placed on either side. The retainers can be fashioned from short sections of tubing.

The LED can be installed in a hole bored into a circular bulkhead of insulating plastic. For best results, the bulkhead should be drilled and tapped to receive two set screws, one for the LED and one to secure the bulkhead inside the cylinder.

Whatever mounting technique is employed, be sure to observe the manufacturer's specifications for soldering. Use particular care with plastic encapsulated diodes, since they cannot withstand the same soldering temperatures and times as metal units.

For operation at high-current levels, any of a variety of heat sinks can be employed. Heat sink materials used in conventional semiconductor circuits can be utilized, but modifications may be required to prevent blocking the light output of the LED. Be sure the heat sink makes intimate contact with a large area metal portion of the LED, in most instances the header or the case. Encapsulated diodes cannot be heat sinked to the degree of metal encased units, but limited protection can be had by heat sinking the contact leads themselves. This is possible since the leads make intimate contact with the LED chip.

POWERING LEDS

Relatively low-voltage and low-current levels are required for proper operation of LEDs and, like incandescent lamps, they can be damaged or destroyed if these ratings are exceeded. Frequently it is necessary to employ a resistor in series with an LED, as shown in Fig. 1-16, to keep the forward current at a safe level. A simple formula can be used to calculate the required series resistance.

$$R_s = \frac{V_{in} - V_{LED}}{I} \qquad \text{(Eq. 1-3)}$$

where,

V_{in} is the voltage available to operate the LED,
V_{LED} is the LED forward voltage,
I is the desired operating current.

An example will show how convenient the basic formula can be in determining the value of an LED series resistor. A typical GaAsP red LED has a V_{LED} of 1.8 volts. Assuming the LED is to be driven at a 5.5-volt logic level and a forward current of 20 mA, Equation 1-3 gives a required series resistance of 185 ohms.

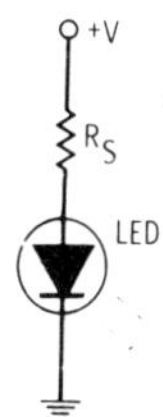

Fig. 1-16. LED current-limiting circuit.

When evaluating or experimenting with LEDs, a convenient method of monitoring forward current is to connect a milliampere meter directly in series with the diode. Since the current flow through all portions of a circuit is equal, the meter current is the LED current. Another approach is to use a series resistor and a voltmeter. According to Ohm's law, current is equal to the quotient of voltage divided by resistance. Therefore, if the series resistor has a value of 10 ohms, the current through the resistor and the LED will be one-tenth of the voltage shown on the meter.

As we shall see later, LEDs can be operated in a pulsed mode for high-power outputs. To avoid damaging the diode, it is important to monitor the peak current through the diode. The series resistor method can be employed in this role, but the voltmeter must be replaced by an oscilloscope. Since pulsed LEDs are usually operated at higher current levels than continuously driven units, it is usually possible to use a 1-ohm or smaller series resistor and still have an easily resolved voltage level. If a 1-ohm resistor is used, voltage across the resistor will equal current through the resistor and LED.

In very high-current pulser circuits, even a 1-ohm resistor can reduce the peak current. Therefore use a 0.1-ohm resistor. Current through the LED and resistor will then be 10 times the voltage shown on the scope. In fast-pulse circuits, be sure to use a noninductive series resistor; otherwise, pulse distortion will occur. A noninductive 0.1-ohm series resistor can be made by connecting ten 1-ohm carbon resistors in parallel. Do not use wirewound resistors in this application, since they possess significant inductance.

OPTICS FOR LEDS

Few applications of visible LEDs require supplemental optics, but infrared LEDs require the use of an external lens when a narrow beam is desired. Subsequent chapters will discuss such applications as optical communications, rangefinders, and intrusion alarms where a collimating lens is almost always necessary.

Unfortunately the integral lens or reflector optical system of most commercial LEDs does not project a very narrow or homogenous beam. Those diodes having a uniform beam structure have relatively wide divergence, and diodes having a narrow beam have a highly structured beam pattern. In fact, a substantial amount of radiation from narrow-beam diodes is usually contained in an off-axis halo surrounding the central beam and originating from light striking the shiny inner walls of the LED mounting package. One very popular, high-quality infrared LED projects only 25% of its total radiant power, available at the lens, into a central beam. The remaining 75% is contained within the off-axis halo surrounding the central beam.

The accompanying series of infrared photographs illustrates the problem of beam homogenity. The photos were made by mounting an LED about a centimeter away from high speed infrared film and activating the LED for a small fraction of a second at 10 mA forward current. The exposures were made in total darkness to avoid fogging the film with stray light.

As shown in Fig. 1-17, the RCA SG1004 has good beam homogenity. This is due to the use of a miniature parabolic reflector to capture the edge emission from the diode chip. The SSL-55CF (Fig. 1-18) also shows a very uniform emission pattern and, again, this is due to the use of a miniature reflector. The border of the infrared pattern

Fig. 1-17. SG1004 infrared beam pattern.

formed by the SSL-55CF appears more defined than that of the SG1004 diode because of the presence of a protective can with a glass-windowed aperture.

The use of a focused glass or plastic lens results in a more structured beam. In Fig. 1-19, the pattern from a lensed version of the SSL-55C is shown. Notice the presence of a distinct halo containing what appears to be a significant portion of the infrared output of the diode. This halo has a very wide divergence and is therefore difficult to collect in many applications. Since most LEDs are rated for power output at the lens opening, the power figure is frequently higher than can be utilized in practical applications.

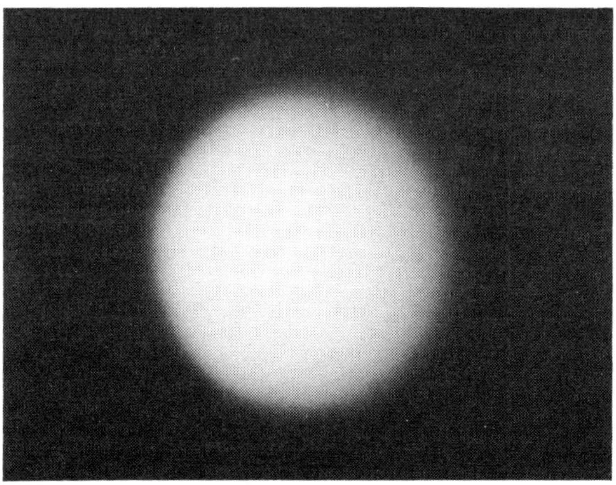

Fig. 1-18. SSL-55CF infrared beam pattern.

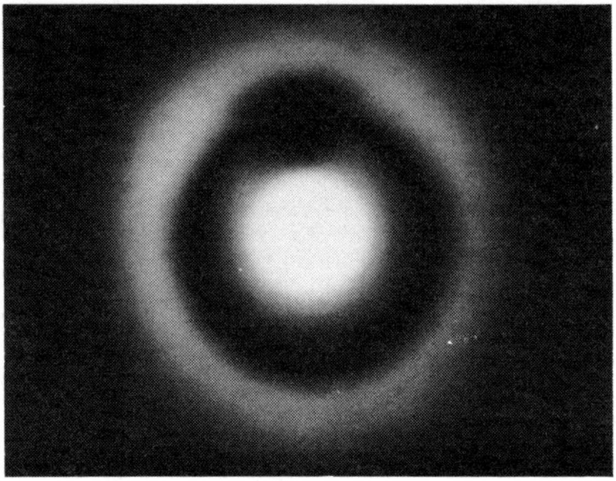

Fig. 1-19. SSL-55C infrared beam pattern.

Another beam pattern of a glass-lensed TIL31 infrared emitter is shown in Fig. 1-20. This device also produces a highly structured beam pattern with a halo. The origin of the halo in Figs. 1-19 and 1-20 is the shiny inner wall of the metal can mounting the lens and enclosing the diode. As shown in Fig. 1-21, some of the radiation from the diode strikes the walls of the can and is reflected to the lens. Since the can is circular, a halo is formed, and since the apparent source size is very large, the halo has a very large divergence.

Several factors must be considered when choosing an external optical system for an LED. Since the objective in most cases is collimation of the radiation into a beam of relatively narrow divergence, it is

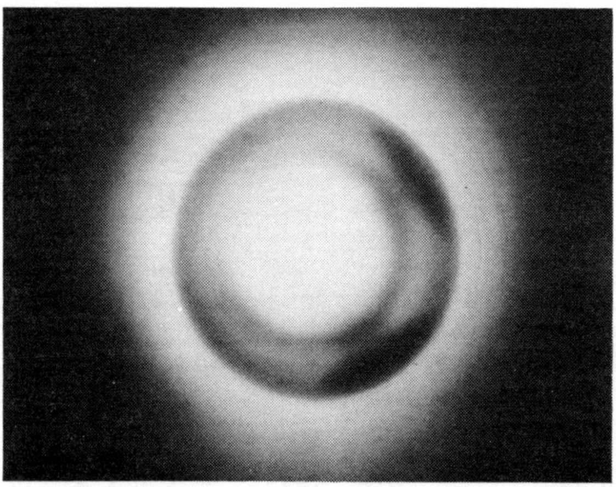

Fig. 1-20. TIL 31 infrared beam pattern.

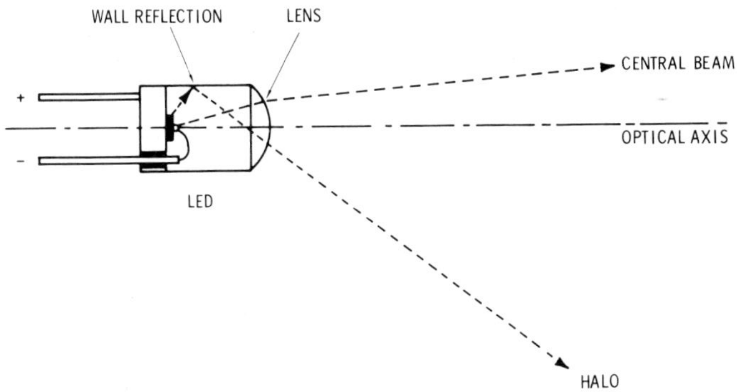

Fig. 1-21. LED halo formation.

desirable to place the lens as far as practical from the diode. The relationship of beam divergence in radians to the LED dimensions and the lens focal length is expressed by

$$\theta = \frac{d}{f} \qquad \text{(Eq. 1-4)}$$

where,

θ is the divergence in radians,
d is the diameter of the source,
f is the focal length of the lens.

Equation 1-4 is a very simple relationship in theory, but in practice it can prove tricky. This is because the result is only as accurate as the values for d and f. Both these parameters can sometimes be difficult to measure. Focal length, for example, is commonly expressed in terms of visible light, but near infrared wavelengths cause an *increase* in this value. Diameter of the source must also be measured with care. In the case of an LED with a self-contained glass lens, the lens, and not the LED chip becomes the source. Since the lens may be several millimeters in diameter and the LED chip only a fraction of a millimeter, the potential error in calculating θ is obvious. These same considerations hold for an encapsulated LED; the encapsulant, not the tiny LED chip, becomes the source.

We have now established the desirability of using a long focal length lens for best results in collimating the radiation from an LED. Unfortunately, it is equally desirable to place the lens as close as possible to the LED in order to collect as much of its radiation as possible. The most practical solution to this apparent dilemma is to use an LED with a very small source size and an f/1 lens. The f/1 lens will permit high collection efficiency, and the small source size will give a relatively narrow divergence.

An examination of an actual system which employs an LED and external lens will illustrate the application of Equation 1-4. The LED is a SSL-55CF, a high-power device with a flat glass window, instead of a self-contained glass lens. This LED incorporates a square chip measuring 0.5 mm on a side, but the source size is enlarged by the internal reflector in which the chip is installed and a small drop of epoxy which covers both chip and reflector. The source size varies slightly from diode to diode due to variations in the diameter of the encapsulant, but is typically about 1.5 mm in diameter.

Some of the radiation from the LED is collected by an f1.2 lens with a focal length of 14 mm. The lens was moved to within 11.5 mm of the LED chip in order to collect as much of the emitted radiation as possible, while still preserving a reasonably small beam divergence. Plugging these values into Equation 1-4 gives a calculated beam divergence of 132 milliradians. The actual divergnce was determined by placing the lens and LED assembly 1 meter from a white card and measuring the diameter of the central beam with the help of an image converter and metric scale. The result was a divergence of 130 milliradians—certainly a close agreement to the predicted value.

Throughout this discussion, the importance of placing the collimating lens close to the LED has been noted. Since LEDs project beams with a relatively wide divergence, more radiation is collected when the lens is moved closer to the diode. Even with a relatively well designed lens system, such as the one just described, a substantial amount of the radiation emitted from the LED window never reaches the lens. In fact, in this particular system only 21.7 mW of the 53.8 mW emitted by the LED was collected and projected by the lens. Worse yet, much of this radiation was projected as an off-axis halo surrounding a relatively narrow central beam. The halo was caused by radiation striking the shiny wall of the LED housing to form a secondary "source." Only 10.5 mW of the original 53.8 mW was contained within the central beam.

Certain other lens-LED combinations will give a higher collection efficiency than this one. But the magnitude of the problem is amply demonstrated by the fact that only about half of the radiation emitted by the SSL-55CF chip even emerges from the glass window. The rest is lost inside the package.

Collection efficiency can be improved by employing parabolic reflectors instead of lenses. This technique has been used with good results for several years. Unfortunately, reflectors are not practical in all systems, since a relatively large diameter is necessary for small beam divergence. Nevertheless, a parabolic reflector can collect far more radiation from an LED than a lens.

CHAPTER 2

LED Circuits and Applications

Because of their low-operating current and solid-state reliability, LEDs are finding use in literally hundreds of circuits and applications. Subsequent chapters will detail the use of LEDs in digital displays, source/sensor pairs, detection systems, and optical communicators. This chapter will describe a variety of circuits which can find use in many applications.

Space limitations prevent listing all possible circuits and applications, but the information presented here should be sufficient for the conception and design of new circuits. Also, electronics experimenter and trade magazines publish a variety of new LED circuits and projects on a regular basis.

INSTALLING LEDS

Use care when soldering an LED into a circuit. Since many LEDs incorporate epoxy construction or encapsulation, excessive heat can quickly destroy a working device. Also, excessive heat can cause LED chips and electrodes to become unsoldered.

Manufacturers' specification sheets always specify the correct soldering temperature, but follow these rules as a general guideline:

1. Use a low-wattage soldering iron
2. Solder rapidly and neatly
3. *Never* solder to the threaded stud or heat sink of a large LED
4. Avoid touching the soldering iron to the optical surface of plastic LEDs.

Some LEDs are made especially for mass production installation. Many can be dip soldered, while others, such as the one shown in Fig. 7-1, are designed for wire-wrap assembly. Some diodes that are excessively heat sensitive can be connected to a circuit board with conductive epoxy.

OPERATING HINTS

For best results, never operate an LED above its maximum ratings. This is particularly true in regard to power dissipation—excessive LED current will cause overheating, resultant degradation, and possibly destruction of a diode.

The best procedure to prevent exceeding power ratings when operating LEDs is to monitor the forward current through the device. Forward current can be monitored by placing an LED directly in series

Courtesy Hewlett-Packard

Fig. 2-1. Visible LED designed for wire-wrap assembly.

with a current meter, or a resistor-voltmeter monitor can be used. In the latter approach, a resistor of known value is placed in series with the LED, and a voltmeter is connected across its leads. If the resistor has a value of 10 ohms, current is equal to one-tenth of the voltage read across the resistor.

High average current density in an LED can result in gradual degradation, or even total destruction, of the device. This effect can take two forms, and both are equally effective in ruining an LED.

The first kind of destruction results from passing a current through the LED that exceeds the specifications of the wire bond connected to the n side of the junction. When this occurs, the bond can overheat and permit the connection point to separate.

The second kind of device destruction has the same result as the first, but through a different mechanism. In this situation, an excessive flow of current literally destroys the LED chip through overheating. The result, when viewed through a 50-power microscope, is an object that more closely resembles a burnt cinder than an LED chip.

Finally, be sure to observe temperature conditions during LED operation. The efficiency of LEDs drops as temperature rises. Under most normal operating conditions, an LED should feel no more than slightly warm to the touch and never hot. If a diode becomes warm or hot during operation, degradation will likely occur. To preclude this possibility, reduce the forward current, or mount the diode on an appropriate heat sink.

VARIABLE-BRIGHTNESS LIGHT SOURCE

The simple circuit of Fig. 2-2 permits the light output from any kind of LED to be varied from an off state to full brightness. The variable resistor provides the brightness control, while the fixed resistor limits the LED current. The circuit has practical application as an indicator or illuminator, and it also serves to illustrate LED operation and current limiting.

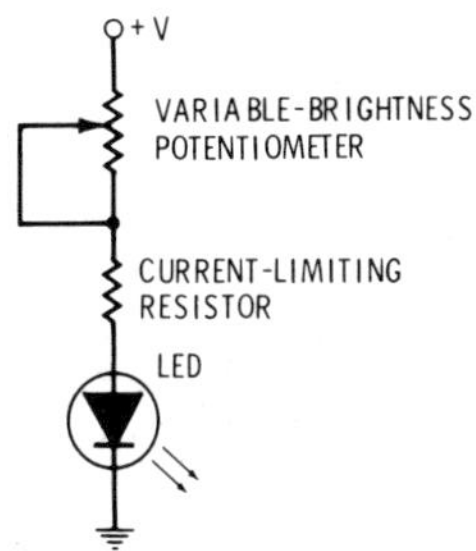

Fig. 2-2. Variable-brightness light source.

Because of the low resistance of LEDs, this circuit and most others can be used to operate two or more light sources. The diodes are simply connected in series.

Variable brightness sources such as this can also be used in physiological tests of human vision and the evaluation of sensitive detection equipment in the presence of very low light levels. The basic circuit can also be used to generate a forward current/optical output graph to determine the linearity of LED output and the maximum operating current. A modified circuit for this application is described next.

LED EVALUATION CIRCUIT

A modified version of the circuit shown in Fig. 2-2 can be used to measure the linearity of LED optical output to an applied forward current. The circuit is shown in Fig. 2-3. The circuit incorporates a current meter in series with the LED to permit accurate monitoring of the LED current. Alternatively, a voltmeter can be connected across the 10-ohm resistor, and current will then equal one-tenth of the voltage reading. Another possibility is to use a variable-current power supply containing its own current meter.

To measure the output of the LED, a silicon solar cell is connected to a current meter. The low resistance of the meter causes the solar cell to be connected to a near short circuit. Therefore, it will operate in an approximately linear manner.

Evaluation of the LED is straightforward. Simply adjust the LED current upward in 5- or 10-mA increments while recording the solar cell output. During the test, the LED should be placed adjacent to

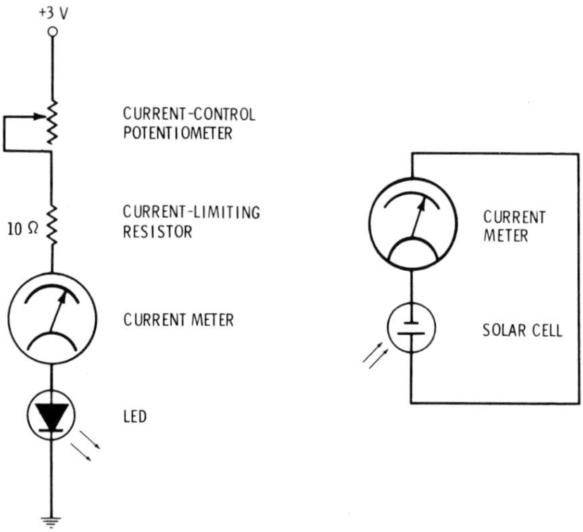

Fig. 2-3. LED evaluation circuit.

the solar cell and the ambient light subdued. If the ambient light level changes or if the LED moves in relation to the solar cell, the output readings will be erroneous.

The results for a typical LED, measured using this technique, are shown in Fig. 2-4. Note the linearity of the curve until the current level becomes too high. The alteration in the output as the current becomes excessive is similar to the curve shown in Fig. 1-8 in Chapter 1.

When conducting this test, be sure to avoid applying too much LED current. The manufacturer's guidelines should be followed if possible. If it is necessary to test the maximum possible operating point for a particular diode, just plot the data for LED input current and solar cell output current as the LED output is increased. As soon as the curve becomes nonlinear, stop increasing the forward current through the LED.

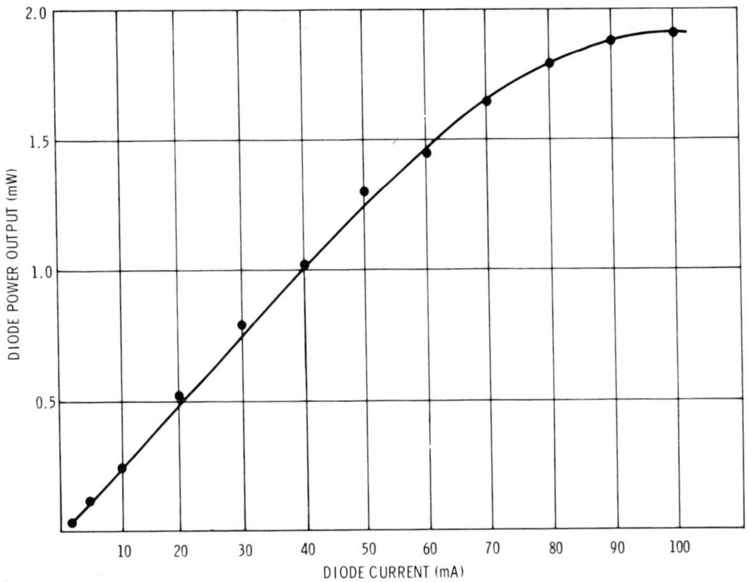

Fig. 2-4. LED power measurement made with circuit in Fig. 2-3.

CONSTANT-BRIGHTNESS LIGHT SOURCES

The light output of an LED is temperature dependent. As temperature decreases, quantum efficiency increases, with a resultant boost in light output. As temperature rises, quantum efficiency and light output drop. By means of a feedback circuit, it is possible to maintain a constant current level through an LED and, therefore, a constant light output. Constant-brightness sources are useful for safe operation of LEDs at absolute maximum ratings in a changing environment, as reference sources, and as calibration sources.

The feedback necessary to regulate the forward current through an LED can be supplied by simple transistor regulator circuits. These circuits have an accuracy of several percent, but better results can sometimes be gained by using optical feedback from the LED. Unfortunately, this method is tricky, since a detector must be optically coupled to some of the LED output and this will frequently block a portion of the LED. Also, if the optical coupling technique is not sturdy, slight movements will cause regulation errors. For these reasons, only transistorized regulators will be described here.

Voltage Regulation

The General Electric *Solid State Lamp Manual,* Part II, presents two constant-current regulator circuits for LEDs on pages 26-27.

These circuits, which are reproduced in Fig. 2-5, provide 4% current regulation over significant variations of the power-supply voltage. The circuits are effective for an operating voltage ranging from 3 volts to the breakdown voltage of the transistors employed. Temperature regulation is 0.3% per degree Celsius.

The R1 of each circuit should be selected for proper biasing of transistor Q2. It is suggested that R1 be equal to 0.6 volt divided by the LED forward-current rating. If the LED is to be operated at a constant current of 50 mA, for example, then R1 should be a 12-ohm resistor. Divide the LED current by 10 to find the value for R2. Therefore a, 50-mA LED current would require that R2 by 0.005 ohms. This small value can be ignored, and R2 can be eliminated in this case.

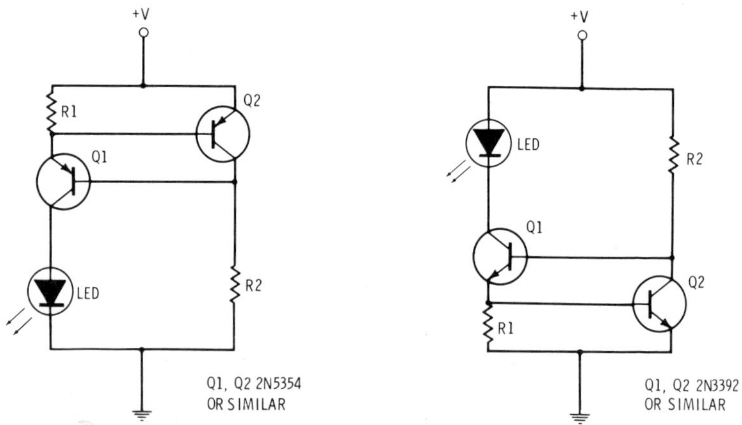

Fig. 2-5. Constant-current operation of LEDs.

Operation of the circuit is straightforward. In the current-source mode, for example, the operating current travels through the LED, Q1, and R1 to ground, since Q2 is *off*. When the voltage exceeds a level determined by R1, Q2 receives base bias and is turned *on*. Excess current is then shunted through Q2 to ground.

The feedback nature of these two current regulators can easily lead to high-frequency oscillation if the gain of the two transistors is sufficiently high. Therefore, careful wiring procedures may be necessary to avoid stray capacitance effects that can sustain oscillations.

Each of these circuits permits the LED to be switched at current levels significantly below the LED forward current by varying R2. With a high value of R2, for example, Q1 will be biased *off*, and no forward current will flow through the LED. Other techniques for amplifying the current through an LED are described in Chapter 5.

Temperature Regulation

By means of a temperature-sensitive thermistor, a high degree of temperature stability can be given to an LED circuit. In the circuit shown in Fig. 2-6, for example, the current regulation is ±3% for a temperature range of from 10° to 50° Celsius.

The circuit employs a feedback effect to achieve current regulation. As the resistance of the thermistor increases because of a decrease in temperature, Q1 receives less bias current, and the LED forward current is reduced accordingly. Since the LED is more efficient at low temperatures, it emits a constant light output.

When the temperature rises, the thermistor responds with a decrease in resistance. This applies more bias current to Q1, and the LED forward current is increased. Again, since LED efficiency falls as temperature rises, the light output remains constant.

For more details about the circuit, see the application note by William Otsuka, "Constant Brightness Light Source," *GaAs LITE Tips,* Vol. I, Monsanto Electronic Special Products, February 1970, page 1.

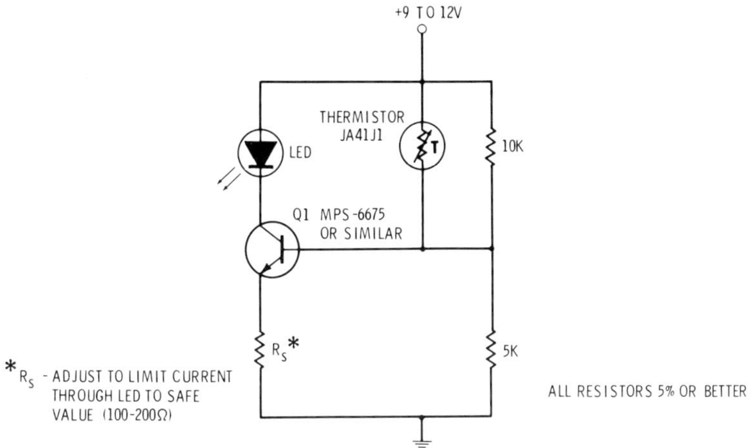

Fig. 2-6. Temperature regulation for constant-brightness LED.

DC POLARITY INDICATOR

A very useful and inexpensive polarity indicator can be made with two red LEDs and a single series resistor. The circuit is shown in Fig. 2-7. Since LEDs possess the same rectification properties as conventional diodes, only the forward-biased diode will light when the voltage being checked is positive. Similarly, when the voltage is negative the reverse-biased diode will not light, but the second diode will.

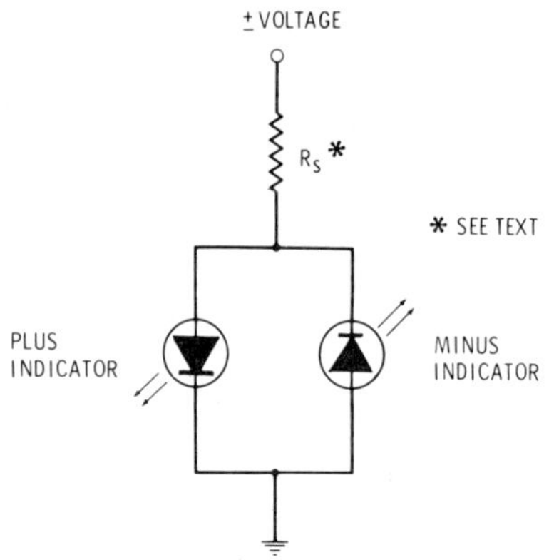

Fig. 2-7. Dc polarity indicator.

The circuit can be simplified by using only one LED, connected to turn *on* when the voltage is positive and to stay *off* when it is negative. In any event, select a series resistor that limits the LED current to a safe value. If a 50-mA LED is used and a 6-volt source is being checked, the resistor should have a value of 120 ohms. If a large variety of voltages are to be checked, use an adjustable series resistor and calibrate it in 1-volt increments.

Recently, LEDs containing two internal chips, connected as shown in Fig. 2-7, have become available. These bipolar LEDs are described in detail later in this chapter.

VOLTAGE MONITOR

Zener diodes and LEDs can be combined to make a variety of LED voltage monitors. Motorola recommends the circuit shown in Fig. 2-8. The circuit monitors an incoming voltage and causes an LED to begin turning on when the input voltage reaches 4.7 volts. When the voltage

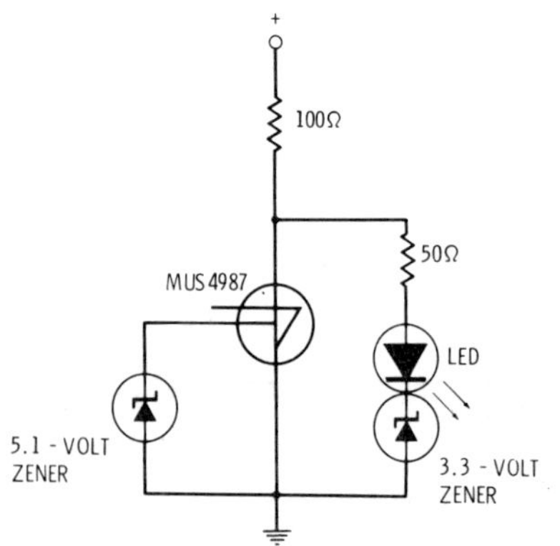

Fig. 2-8. LED voltage monitor.

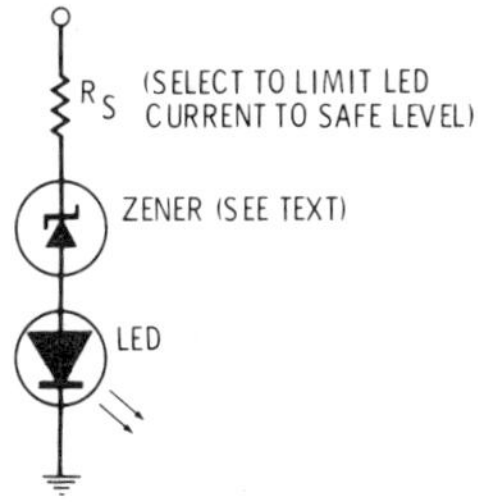

Fig. 2-9. LED voltage-level indicator.

exceeds 6.9 volts, the LED is extinguished as the voltage is shunted through the silicon unidirectional switch and the 5.1-volt zener diode.

The circuit is handy for visually monitoring the power-supply voltage for logic circuits requiring 5.5 volts. So long as the LED is on, the supply is delivering between 4.7 and 6.9 volts. The relative brightness of the LED provides a subjective indication of the actual voltage. For more information about the circuit, see *Optoelectronics at Work,* a Motorola LED application brochure.

An even simpler circuit employs an LED and a zener diode to indicate when a voltage has reached a desired level. The circuit is shown in Fig. 2-9. In operation, no current flows through the LED until the breakdown point of the zener diode is reached. The LED then begins to turn *on* and become brighter as the voltage is increased.

Select a zener to give the desired voltage indication. For example, with a 6.2-volt zener, the LED will begin to glow noticeably with an input of about 5.8 volts.

LOGIC STATUS INDICATOR

A very popular use for visible LEDs is logic status indication. An LED is simply connected in a logic line, usually with a series resistor, and it lights up when the logic device is in the *on* state. Fig. 2-10 shows a typical circuit diagram. The series resistor should be selected to prevent the maximum current rating of the LED from being exceeded.

LOGIC TESTER PROBE

A visible LED and a few other components can be used to make a simple logic status indicator probe. The circuit for the probe is

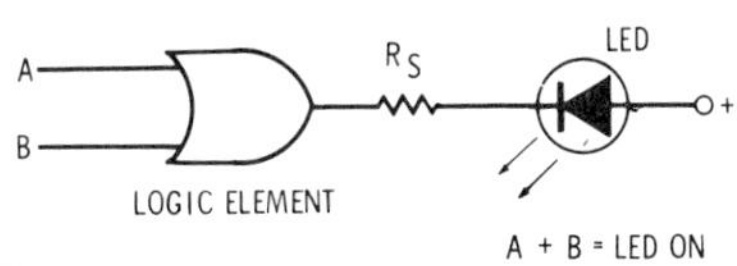

Fig. 2-10. Logic status indicator.

shown in Fig. 2-11. In operation, the probe is connected to the positive and ground terminals of the logic circuit undergoing test. A probe connected to the 20K resistor is then used to check the logic levels of various parts of the circuit under test. The circuit has a high input impedance and will respond to a minimum logic level of about 1.5 volts. A high logic level will turn the LED on, while a low logic level will keep it off.

For maximum utility, the logic tester probe should be installed in a self-contained housing. A small, plastic case will do. Two clip leads should be provided for the positive and ground connections, and a standard probe should be adapted as the test probe. Ideally, mount the

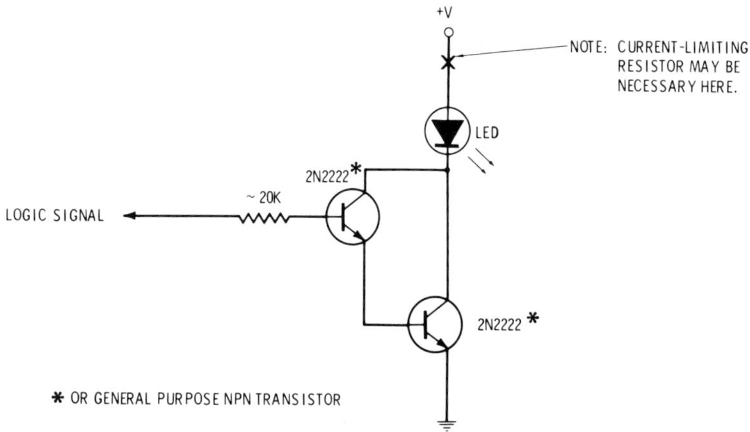

Fig. 2-11. LED logic probe.

probe in one end of the box and the LED in the other, to permit maximum convenience.

MULTIPLE-COLOR LEDS

Two kinds of multiple-color LEDs are available. One uses a single chip of gallium phosphide (GaP) to give a simultaneous green and red emission from each side of the junction. Since the red emission saturates at relatively low-current densities, the diode can emit hues ranging from red to green as the forward voltage is increased. A second approach to multiple-color output is provided by a two-chip LED. In this device, a separate GaP chip is provided for red and green emission.

Multiple-color LEDs are current-sensitive devices and can therefore be used to visually indicate a current or voltage by means of their subjective color emission. The most basic circuit would simply employ the LED and a current-limiting series resistor and would indicate the relatively low voltages required for the LED (e.g., 1.6 volt for red and 3.5 volts for green). The LED could be calibrated for its various colors by connecting a voltmeter to the power supply while varying the voltage and observing the respective LED hues.

Zener-diode arrangements can be used to measure higher voltages. Since the zener will not conduct until its breakdown voltage is reached, an LED in series will remain off until breakdown. The LED hues would have to be recalibrated for the higher-voltage situation, but the circuit would provide a quick indication of voltage by color variations.

There are other unique circuit possibilities for multiple-color LEDs. Optical multiplexing is one, since a single optical fiber can be used to transmit all of the various wavelengths of the device. Wavelength-selective filters, at a fanout of fibers on the receive end, would decode the multicolored signals. Other applications include overload indication, logic-level indicators, battery level indicators (green means correct voltage, while red signals the need for battery replacement), and even a simple indicator for some applications that use an analog meter. In the latter case, the recognizable hues of the LED could represent fuel-tank capacity, time intervals, battery charging, temperature, etc.

BIPOLAR LEDS

In 1973, the first commercially available ac-dc solid-state lamp was introduced—the MV5094. This device is simply a standard epoxy-encapsulated package containing two, back to back, GaAsP red-

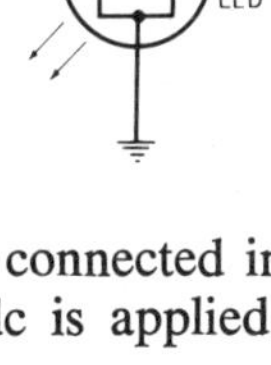

Fig. 2-12. Bipolar LED circuit.

emitting LEDs. Since the two diodes are connected in a reverse-parallel configuration, one will light when dc is applied, and both will light when ac is applied.

The MV5094 can be operated from ac voltages ranging up to 110 volts by means of a suitable series resistor. A circuit for this mode of operating, showing the bipolar LED and series resistor, is presented in Fig. 2-12. Table 2-1 gives the resistor values required to operate

Table 2-1. Resistor Values for Bipolar LED

Voltage (rms)	Series Resistor
5.0	360 ohms, ⅛ W
6.3	470 ohms, ⅛ W
9.0	750 ohms, ⅛ W
12.0	1000 ohms, ⅛ W
15.0	1300 ohms, ¼ W
18.0	1600 ohms, ¼ W
24.0	2200 ohms, ¼ W
28.0	2700 ohms, ½ W
48.0	4700 ohms, ½ W
110.0	11,000 ohms, 2 W

NOTE: These values apply for LED forward current of 10 mA and forward voltage of 1.56 V.

the diode at a forward current of 10 mA from a variety of ac voltages. Series resistance values for other ac voltages can be calculated from the following formula:

$$R_s = \frac{V_{rms} - V_f}{I} \qquad \text{(Eq. 2-1)}$$

where,

V_{rms} is the available ac voltage,
V_f is the LED forward voltage,
I is the desired LED forward current.

A bicolor tri-state LED, the MV5491, is also available. This device is identical to the MV5094, except one of the red GaAsP LEDs has been replaced by a green GaP LED. By altering the bias current through each of the LEDs, the MV5491 can be made to emit various hues ranging from red to green.

The dual color output of the MV5491 makes it a true polarity indicator. When connected to a dc source, either the red or green chip will be lighted, thus supplying a positive indication of polarity.

Since the GaAsP and GaP LEDs each have different voltage-current requirements, a biasing network is necessary to achieve equivalent illumination from both chips. The situation is complicated by the fact GaP emits a range of wavelengths in the green near the peak of the photopic luminosity curve, while GaAsP emits in the red with a photopic luminosity of only about 10%.

To facilitate matching the intensities from both the red and green chips, Monsanto suggests the simple circuit in Fig. 2-13. The circuit is designed to operate from a 5-volt logic level, but other power-supply voltages can be employed by altering the values of R1 and R2. Two formulas are required to calculate the values of R1 and R2. The total value of R1 + R2 is given by:

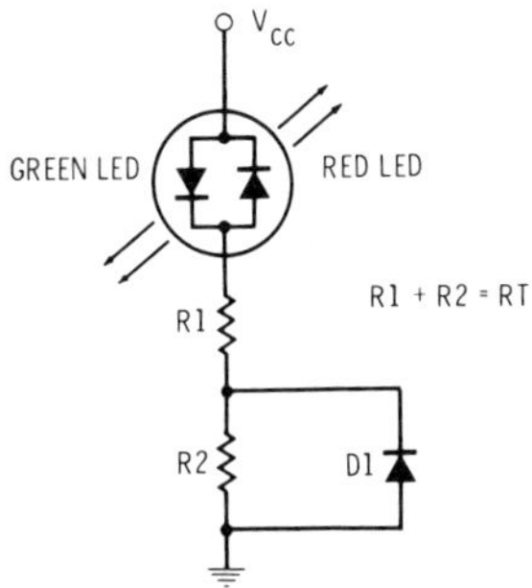

Fig. 2-13. Tri-state LED circuit.

$$R_T = \frac{V_{cc} - V_{RLED}}{I_{RLED}} \qquad \text{(Eq. 2-2)}$$

where,

V_{cc} is the power supply voltage,
V_{RLED} is the forward voltage of the red LED,
I_{RLED} is the desired forward current of the red LED.

R1 is given by:

$$R1 = \frac{V_{cc} - (V_{GLED} + V_D)}{I_{GLED}} \qquad \text{(Eq. 2-3)}$$

where,

V_{GLED} is the forward voltage of the green LED,
V_D is the forward voltage of the 1N914 diode,
I_{GLED} is the desired forward current of the green LED.

By means of Equations 2-2 and 2-3, a table can be prepared to permit the matching of the red and green LEDs over a range of currents. Table 2-2 is an example. To read the table, select the desired current through both the red and green diodes and proceed across and down the table until the required values of R1 and R2 are reached. For other current values, use Equations 2-2 and 2-3. Note that not all current combinations are possible. This is because the equations give a *negative* resistance for R2 in these cases, an impossible situation in the circuit shown in Fig. 2-13.

Both the MV5094 and MV5491 have a wide range of novel applications. The MV5491, for example, can be used to indicate the presence of undesirable negative undershoot in pulse circuits. In this role the device would be connected directly to the output of the pulse circuit. Depending on the connection polarity, either the red or green chip should glow with each applied pulse. Any illumination from the other chip indicates undershoot. The MV5491 might also find application in phase detectors.

Table 2-2. Resistor Combinations for Tri-State LEDs

	Green														
	10 mA			20 mA			30 mA			40 mA			50 mA		
Red	RT	R1	R2	RT	R1	R2	RT	R1	R2	RT	R1	R2	RT	R1	R2
10 mA	344	230	114	344	102	242	344	63	281	344	44	300	344	34	310
20 mA				170	102	68	170	63	107	170	44	126	170	34	136
30 mA				112	102	10	112	63	49	112	44	68	112	34	78
40 mA							84	63	21	84	44	40	84	34	50
50 mA							67	63	4	67	44	23	67	34	33
60 mA										55	44	11	55	34	21
70 mA										47	44	3	47	34	13

NOTE: Resistor values valid for power-supply voltage of 5 V. All values in ohms.

If either of these self-contained bicolor LEDs are unavailable, individual LEDs can be connected back-to-back to achieve similar results. This configuration is not as convenient as the self-contained approach, but it will suffice in many applications. The same formulas and tables for calculating series resistance for self-contained bipolar LEDs apply.

LED FLASHER

A very simple visible or infrared flasher can be made with a low-cost unijunction transistor, a capacitor, and a few resistors. A representative circuit is shown in Fig. 2-14. The UJT forms the heart of a relaxation oscillator, having a frequency that can be altered by varying

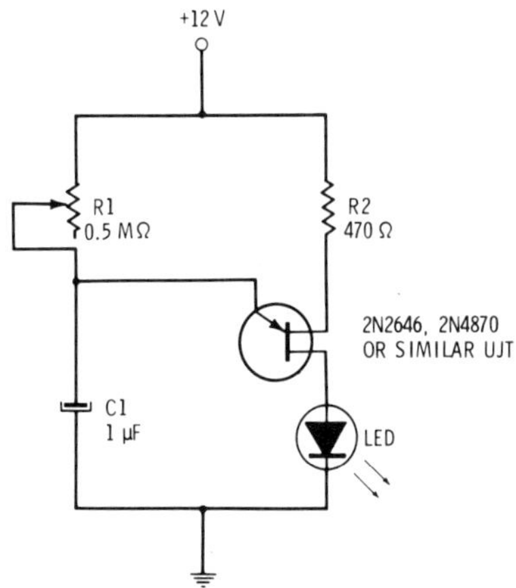

Fig. 2-14. LED flasher.

R1 and C1. The circuit pulse width is controlled by C1; a value of 1 microfarad will give a width of about 20 milliseconds. Adjusting R1 will give a frequency ranging from a pulse every few seconds, to more than 200 pps.

NEGATIVE-RESISTANCE LED CIRCUITS

The GND-50G is a negative-resistance, pnpn, GaAs:Si LED. It is coated with a green phosphor unconverter and has a peak output at 540 nm. The diode is available in the United States from Shigoto Industries Ltd.

The GND-50G has very unique optoelectronic properties. Being a four-layer device, no current flows when a low voltage is applied. However, when the applied voltage exceeds the breakdown voltage of the diode (15 to 30 volts), the device goes into conduction. The GND-50G acts like an LED with a discrete threshold voltage.

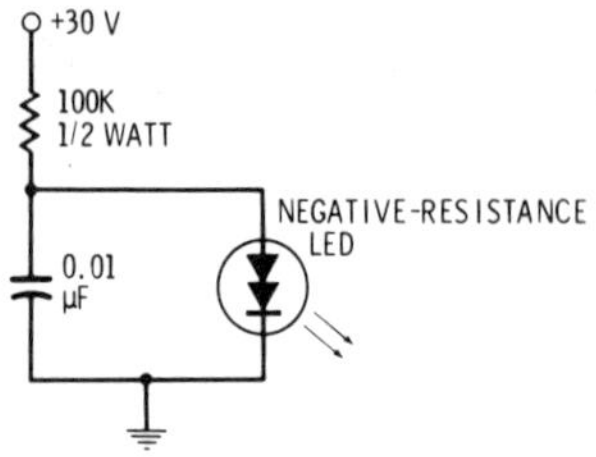

Fig. 2-15. Negative-resistance LED pulse generator.

The negative-resistance characteristic of the GND-50G can be used to make a simple relaxation oscillator, and a suggested circuit is shown in Fig. 2-15. In operation, a voltage charges a capacitor until the breakdown voltage of the LED is reached. At that point, the LED conducts and allows the capacitor voltage to be shunted to ground. As the capacitor discharges, the LED emits a pulse of light with an amplitude and width determined by the value of the capacitor. The repetition rate of the circuit can be varied by adjusting the charging resistor.

PROXIMITY SENSOR

A simple proximity detector can be made with an LED and a single field-effect transistor. The circuit is shown in Fig. 2-16. Normally, a current flows from the source to drain of the FET and biases the LED, since the gate is unbiased. When the gate lead is touched the FET is turned all, or partially, *off* and the LED responds accordingly.

The circuit can be made very sensitive in a dry environment by attaching a short lead to the FET gate and placing it near a position to be observed. Normally, the LED will be *on,* but when a human body walks past the sensor wire, the LED will momentarily turn *on* and *off.* A system built by the author responded to a hand, at a distance of 0.5 meter away, in a very low-humidity environment.

LED TEMPERATURE SENSOR

LEDs are temperature-sensitive devices; as temperature decreases, efficiency increases and wavelength decreases. By powering an LED

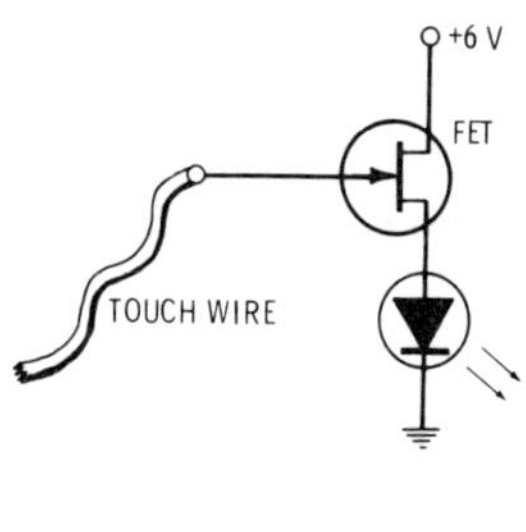

Fig. 2-16. LED proximity indicator.

at a steady current of, for example, 50 mA, it is possible to devise a simple temperature-indicating device with a remote capability. A suggested circuit is shown in Fig. 2-17.

In operation, the LED derives power from a constant-current power source and projects a radiant beam of infrared radiation to the photodiode. The current meter which is in series with the photodiode then indicates a current flow which is directly proportional to the amplitude of the radiation striking the photosensitive surface of the photodiode.

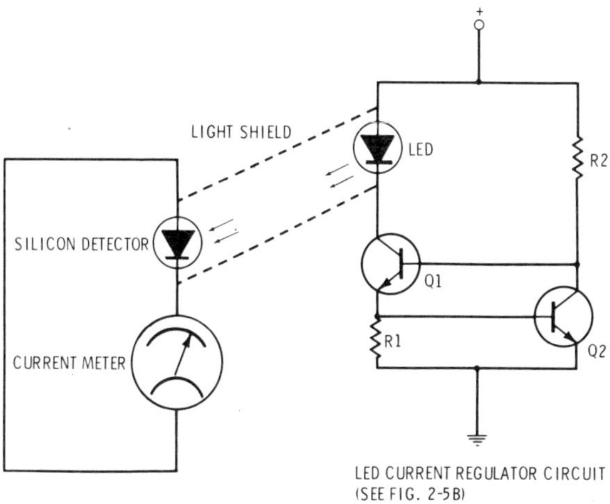

Fig. 2-17. LED temperature indicator.

After the system is assembled, it must be calibrated. This is accomplished by placing a thermometer adjacent to the LED and exposing both to several discrete temperatures. The alignment between the LED and detector must be fixed during and after calibration and this can be arranged by means of fiber optics or mounting both the LED and detector to a rigid surface. If the fiber optic technique is used or if the LED and detector are mounted to a rigid surface which is portable, calibration can be facilitated by placing the LED portion of the assembly in the draft from an air conditioner, and then in the draft from a furnace. This will give several calibration points and permit the meter to be calibrated.

Since this system is steady state, therefore, it is not immune to the ambient light striking the detector. If the temperature sensor is to be operated in the presence of ambient light, a pulsed LED and an ac-coupled receiver can be employed. Several pulsed LED transmitter and receiver circuits are described in this and subsequent chapters.

HIGH-CURRENT PULSER

A simple two-transistor pulse circuit can be employed to obtain high-current pulses from an LED. The circuit is shown in Fig. 2-18. With the values shown, the circuit will drive an SSL-55C, or similar LED, with 2.7-ampere pulses having a width of about 17 microseconds. Repetition rate will be about 120 Hz. Current drain from a small 7-volt mercury battery is only about 5 mA. This gives a useful battery life of more than 32 hours from a Mallory TR-175.

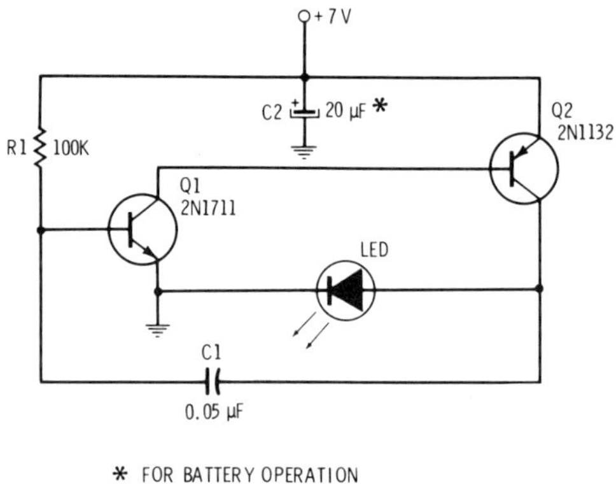

Fig. 2-18. High-current LED pulser.

The circuit operates in the following manner. Referring to the circuit diagram, note that when the LED is *off,* Q2 must also be *off.* Since Q1 controls Q2, Q1 must also be *off.* Resistor R1, however, supplies enough bias to turn Q1 *on,* causing Q2 to saturate. When Q2 is *on,* the LED is connected directly between the plus voltage and ground, with only the highly conductive collector-emitter circuit of Q2 acting as a very small series resistor. This is how the circuit produces a very high current. If the series resistance of the LED and Q2 is 2.2 ohms, and the power supply is set for 6.0 volts, the current will be an impressive 2.7 amperes.

When both transistors and the LED are *on,* capacitor C1 begins to charge until a point is reached where Q1 begins to conduct less, and then Q2 and the LED are turned *off.* The capacitor is no longer being charged, so it discharges through some of the circuit components until its charge is too small to keep Q1 *off.* Then Q1 turns *on,* C1 begins charging, and the cycle repeats itself.

When the circuit is operated from a high-current power supply, the large storage capacitor (C2) across the voltage input will not be needed. The capacitor is needed for battery operation, since the direct short circuit, through the LED and Q2, that occurs when Q2 turns *on,* places an excessive current drain on the battery and significantly alters pulse shape. In fact, the battery voltage will actually drop considerably during individual pulses and cause the repetition rate to increase.

The inverse relationship between battery voltage and repetition rate is a handy indicator of battery replacement time. The normal 120 Hz will increase to several times that value when the battery voltage drops from 6 to about 4 volts.

The basic circuit in Fig. 2-18 is very useful for LED illuminator and transmitter applications. The author has employed the circuit in miniature LED communicators and detection systems with good results, and several applications for the pulser are described in Chapters 5 and 6.

AVALANCHE TRANSISTOR PULSE GENERATOR

A simple avalanche transistor circuit, requiring only five components, will deliver very high-current pulses to an LED. Depending on component values, the current can range up to 15 A or more, the pulse width can range from a few nanoseconds to 75 ns, and the repetition rate can range from 0 to 30 kHz.

A sample circuit is shown in Fig. 2-19. In operation, the capacitor charges through R1 and R2 until the Q1 collector-emitter breakdown voltage is reached. Then, Q1 avalanches, and C1 discharges through Q1 and the LED. When Q1 turns *off,* the cycle repeats itself in a relaxation oscillator mode. The base bias to Q1 is provided by R3.

Many ordinary silicon and germanium transistors will operate in an avalanche mode. Since specially manufactured avalanche transistors are expensive, an acceptable and economical procedure is to build a test circuit with a transistor socket in place of Q1. By inserting various transistors into the socket and adjusting the supply voltage upward until oscillation begins, the breakdown voltage of each transistor can be measured. Not all transistors will work in the circuit, but many will. For best results, try switching transistors such as the 2N918, 2N2222, 2N3643, etc.

A transistor such as the 2N2222 may have a breakdown voltage of 60 volts or more. In a circuit like the one in Fig. 2-19, such a transistor will easily deliver 15-ampere pulses to the LED, with a width of about 50 ns. Good, quality, infrared-emitting GaAs:Si LEDs will generate more than a quarter of a watt of infrared radiation from this much current, but be sure to check the maximum permissable pulse-current ratings before inserting an LED into the circuit; otherwise, device destruction may result.

A fast oscilloscope can be used to measure the peak current of this pulse generator. Connect a 0.1-ohm resistor in series with the LED and connect the scope probe across the resistor. The voltage on the scope will correspond to one-tenth the current through the LED. Circuit pulse width and repetition rate can also be measured in this test.

A scope with a bandwidth of at least 15 MHz is necessary for this test. If a fast scope is not available, improvise by connecting a slower scope across C1 to measure the transistor breakdown voltage. Assuming a typical dynamic impedance of 1.5 ohms for the LED and 2.5 ohms for the transistor, we can calculate the approximate peak current by using Ohm's law. If we assume a transistor with a breakdown voltage of 60 volts and a total discharge circuit resistance of 4 ohms, an approximate current of 15 A will be delivered to the LED.

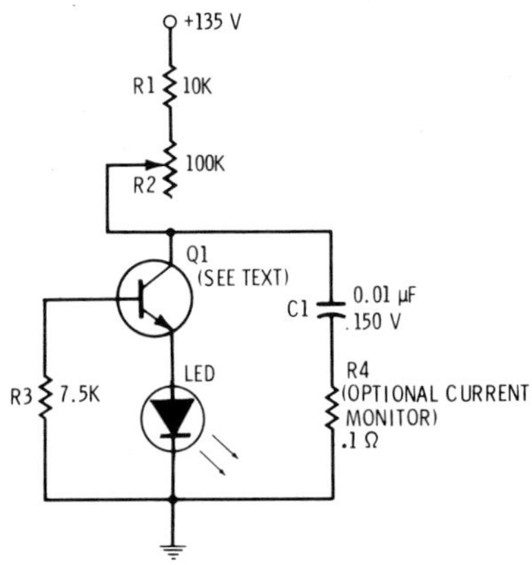

Fig. 2-19. Avalanche transistor pulse generator.

The very fast pulse width and high-current capability of avalanche transistor circuits make them ideal for powering injection laser diodes. By using specially designed avalanche transistors and parallel circuits, containing two or more discharge circuits, more than 100 A can be delivered to a single injection laser.

SOLID-STATE TELEVISION

Recent advances in detector technology have made possible MOS and self-scanned, charged-coupled, one- and two-dimensional detector arrays. These arrays have made possible miniature solid-state television cameras, small enough to fit in a shirt pocket. Advances in LED technology will no doubt permit the fabrication of LED arrays with sufficient resolution for displaying a recognizable video image. Already, a linear array of 48, red-emitting GaAsP LEDs is available. The diodes are mounted on 0.127-mm centers, thereby providing a resolution of nearly 8 lines per millimeter.

A linear array of photodiodes or phototransistors and a similar array of LEDs can be used to obtain crude television pictures. A scan-

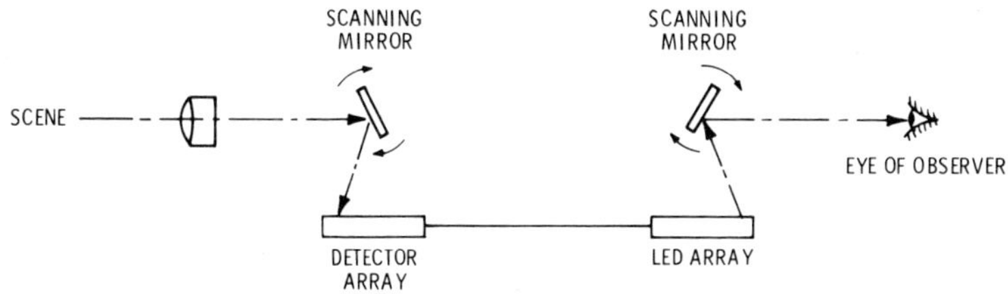

Fig. 2-20. Experimental LED television system.

ning mirror and optical system is required at both arrays to permit a two-dimensional scene to be detected and transmitted to the LED array. A block diagram for such a system is shown in Fig. 2-20.

While this LED television scheme is impractical, it does demonstrate the future possibility of an all solid-state television system. Two-dimensional arrays and processing circuitry will eventually eliminate the need for scanning mirrors.

OTHER APPLICATIONS

In addition to the applications previously described, LEDs can be employed in level indicators, temperature monitors, color comparators, telemetry, position indicators, air-flow sensors, smoke detectors, film recording, and numerous other applications.

Smoke detection is achieved by causing an LED beam to strike a detector after passing through a light-tight tube through which air is blown. The presence of smoke is indicated by a reduced photodetector current.

Film recording employs a voice-modulated, visible LED to generate an optical sound track on movie film. The LED technique requires less power and has better frequency response and linearity than conventional film-recording light sources.

In addition to circuits and devices, LEDs are finding increasing use in computers, electronic calculators, medical instrumentation, measuring instruments, and space hardware. In the near future, LEDs will be commonly employed in telephones, automobiles, home appliances, clocks, annunciators, and remote-controlled toys.

A major application for LEDs is in the fields of optical communication and detection. Numerous applications can be found for an optical communicator. Besides transmitting voice, computer data, or television, the apparatus for a basic optical communicator can easily be adapted for remote control, telemetry, and intrusion detection. Optical detection devices can be used in applications as diverse as detecting automobiles and as infrared guidance devices for the blind.

Pairing an LED with an appropriate detector makes a source/sensor pair with a great range of applications. Source/sensor pairs can be used as optical isolators, gates, transmission sensors, and in other roles.

Subsequent chapters discuss some of these applications in more detail. First, however, some LED detector and receiver circuits will be described.

LED DETECTOR CIRCUITS

Several types of semiconductor detectors can be used to detect the radiation from both visible and infrared LEDs. The most popular detectors include silicon solar cells, phototransistors, photodiodes, and light-sensitive FETs. Even the LED itself can be used to detect the radiation emitted by another LED. Most of the detectors, several of which are shown in Fig. 2-21, have a peak spectral response close to the peak wavelength of GaAs infrared LEDs.

Each of the various detectors has advantages and disadvantages, and no single detector is suitable for all applications. Solar cells are inexpensive, and their large surface area permits operation without an external lens. The latter characteristic is useful in applications where the precise alignment required of optical systems is undesirable. For example, an LED communications receiver employing a lens and small area detector must be very carefully aligned, with respect to the

Courtesy EG&G

Fig. 2-21. Assorted silicon photodetectors.

transmitter, to cause the maximum amount of radiation to fall on the active surface of the detector. The solar cell is a wide-area detector and no such alignment is required. Unfortunately, solar cells are as sensitive to ambient light as LED radiation. Therefore they cannot be used in bright, ambient light without the presence of a filter or light baffle to restrict unwanted light. Solar cells are also relatively slow, with typical response times being several microseconds or more.

Phototransistors are very sensitive and have the important advantage of internal gain. Darlington phototransistors are particularly sensitive. Phototransistors are available in a wide variety of configurations at relatively low cost. For best results, they must be used with an external lens, and this makes alignment with an LED source tricky. Like solar cells, phototransistors have a relatively slow response time, but there are special circuits which can be used to speed up response time. Finally, phototransistors are very sensitive to ambient light due to their high internal gain. For this reason it is essential to use a filter, light baffle, or both when operating a phototransistor in the presence of bright light sources.

Light-sensitive FETs provide the advantages of the FET while ensuring fast response time. Phototransistors have typical response time measured in microseconds, while a light-sensitive FET has a rise time of perhaps 50 ns. Light-sensitive FETs are small-area detectors, so an external lens is necessary for best results. Also, light-sensitive FETs are more expensive than phototransistors.

Photodiodes are very useful in fast pulse width LED systems such as intrusion alarms and pulsed communicators. PiN photodiodes are particularly useful in situations where high ambient light is present, since they produce a photocurrent which is linear over seven decades of incident light. This means a photodiode can be used in the presence of indirect sunlight and still be capable of detecting very weak pulsed radiation from an LED without a filter or light baffle.

Photodiodes are available in both small and large active area configurations. They are more costly than most other detectors, but their desirable operating characteristics usually offset the additional cost.

A special type of photodiode is the avalanche detector. This detector is characterized by internal gain and a resultant high degree of sensitivity. Avalanche detectors, however, require carefully designed biasing circuitry and temperature regulation to provide the precise voltage level necessary for maximum operating efficiency. For this reason they are available as integral detector-amplifier modules containing the required biasing circuitry. Avalanche detectors are among the most expensive detectors available.

Several receiver circuits employing some of these detectors will be described in detail in Chapters 5 and 6. For simpler applications, try the basic circuits which follow.

SOLAR CELL DETECTOR CIRCUIT

A simple circuit for amplifying pulsating signals received by a solar cell is shown in Fig. 2-22. In operation, radiation striking the active surface of the solar cell is converted into a photocurrent. The 2-μF capacitor blocks dc components of the photocurrent (originating from sunlight and other dc sources) and passes the pulsating components. The 2N2907 transistor amplifies the pulsating signal. The output signal can be fed into magnetic phones or further amplified.

The solar cell will require a filter and light baffle if the circuit is operated in sunlight. The 2-μF capacitor will block dc signals from the solar cell, but the solar cell will not detect pulsating radiation if it becomes saturated by excessive sunlight.

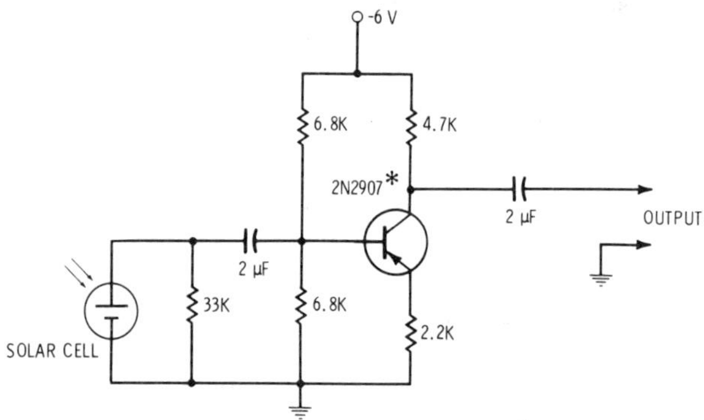

* OR OTHER GENERAL PURPOSE PNP TRANSISTOR

Fig. 2-22. Simple solar-cell receiver.

PHOTOTRANSISTOR CIRCUITS

Phototransistors are among the most economical and sensitive of photodetectors. Unfortunately, they have slow frequency response and tend to saturate in the presence of ambient light. The slow response time of a phototransistor and most other semiconductor detectors is caused by junction capacitance. The capacitor, formed by the junction, must be fully charged before the detector can begin responding to all of a pulse. For this reason, fast pulses are not detected as well as slow pulses.

An easy way to reduce the effect of junction capacitance is to reduce the value of detector load resistance. Since detector time constant is directly proportional to the product of its capacitance and the load resistance, a small value of load resistance gives a faster time constant.

Unfortunately, reducing phototransistor load resistance also reduces sensitivity. This is because a phototransistor is essentially a light-controlled current source. Therefore, the voltage developed across the load resistor is directly proportional to the value of the load resistor.

Many LED pulse circuits require a sensitive detector, with fast response time, and several novel circuits have been devised to enable the phototransistor to provide both. The simplest of these "speed-up circuits," as they are commonly known, is shown in Fig. 2-23. The 2N2907 transistor is connected as a common-base amplifier to provide a low-impedance load to the phototransistor. This satisfies the

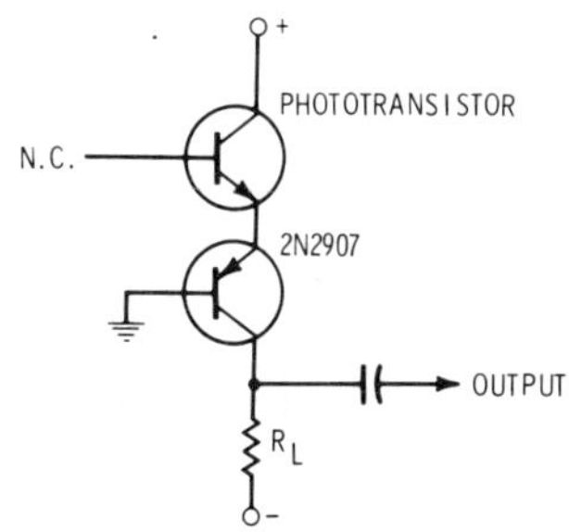

Fig. 2-23. Phototransistor speed-up circuit.

requirement for reducing the time constant of the phototransistor and speeding up its response time. High sensitivity is provided by connecting a load resistor to the output of the 2N2907.

Various configurations of the basic speed-up circuit can be employed. One interesting approach is to replace the common-base transistor amplifier with an integrated-circuit operational amplifier (op amp). One circuit possibility is shown in Fig. 2-24. Gain can be increased at the expense of frequency by increasing the value of feed-

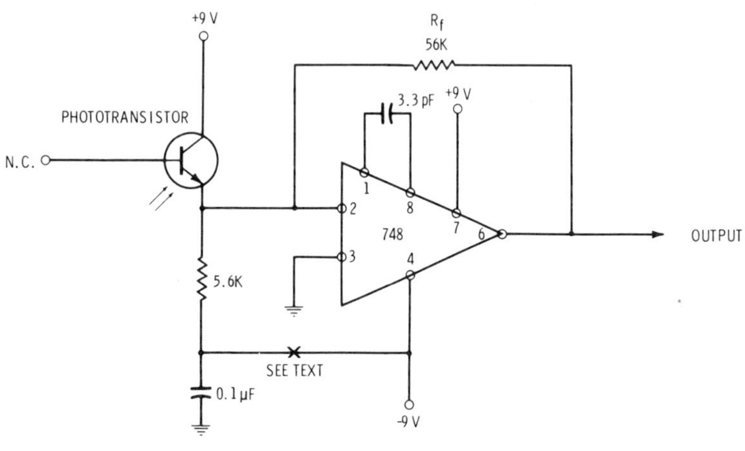

Fig. 2-24. Operational amplifier phototransistor circuit.

back resistor R_f. For high gain and high-frequency response, a better-quality op amp can be substituted for the low-cost 748 shown here.

The op amp speed-up circuit offers several advantages over the simpler transistor speed-up circuit shown in Fig. 2-23, particularly the ease with which gain can be controlled. But since the op amp circuit is direct coupled, it will present an output voltage proportional to *any* light striking the phototransistor. To permit an output voltage from a dc light source (such as the sun) to be cancelled, insert a 15K potentiometer at point "X" in Fig. 2-24. Suitable adjustment of this potentiometer will zero the op amp output.

Many phototransistors do not have a base lead. In effect, light striking the base region provides a base bias. This means very small signal levels can be difficult to detect since there is not sufficient incoming radiation for optimum biasing of the transistor. When a phototransistor circuit is to be operated in the dark and when the input signal is of small magnitude, either substitute a phototransistor with an external base connection or illuminate the phototransistor with a small amount of radiation from an adjustable brightness LED. A suggested circuit is shown in Fig. 2-25. The potentiometer should be set for optimum sensitivity of the phototransistor and readjusted as ambient light changes. A visible or infrared-emitting LED can be employed, but the latter will provide a greater range of sensitivity control.

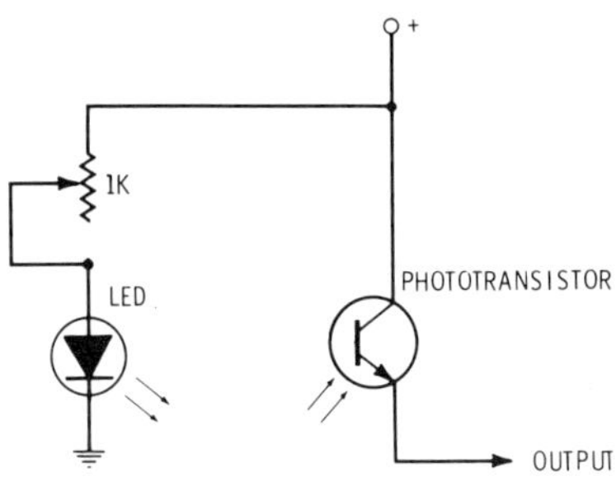

Fig. 2-25. LED bias circuit for phototransistor.

PHOTOTRANSISTOR CONTROL CIRCUIT

A simple method of using a phototransistor to count objects in a break-beam detection system is to connect the transistor directly across the capacitor in a unijunction transistor relaxation oscillator. A sample circuit is shown in Fig. 2-26. In operation, radiation from an infrared LED normally strikes the sensitive surface of the phototransistor. When the infrared is interrupted, however, C1 charges until the UJT switches on and triggers the gate of the SCR. The SCR circuit activates a counter, alarm, or other device.

This basic circuit can be altered to vary its characteristics. Optics for the phototransistor and LED will improve detection range. The

LED can be connected to the receiver power supply (be sure to limit the LED current to a safe value). The circuit can be made immune to momentary transients, such as falling leaves or blowing debris, by adjusting R1, C1, or both so that a preset interruption time is necessary to activate the SCR. With small values for R1 and Cl, the relaxation oscillator will oscillate rapidly, and very brief interruptions will trigger the SCR. But large values for R1 or C1 will slow down the oscillation rate considerably and longer interruption times will be required. By using a 100K potentiometer for R1, the interruption interval can be conveniently altered.

Circuits similar to the one in Fig. 2-26 can be designed for other detectors. For best results, try photodiodes and light-activated, silicon controlled rectifiers (LASCRs).

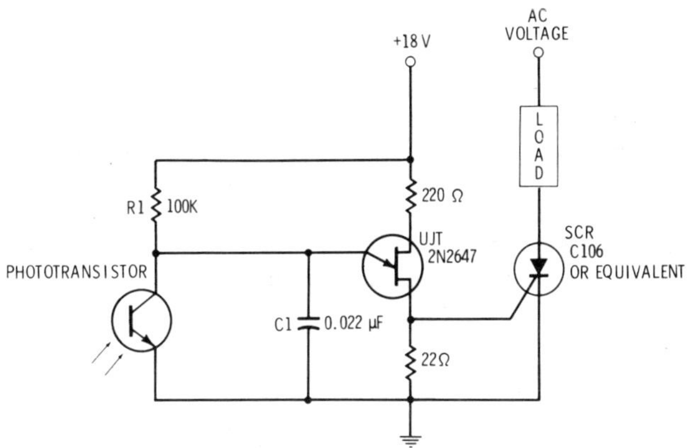

Courtesy General Electric

Fig. 2-26. Phototransistor control circuit.

PHOTODIODE CIRCUITS

Photodiodes can be operated in two basic configurations, the unbiased *photovoltaic mode* and the biased *photoconductive mode*. The photovoltaic mode provides the highest signal-to-noise ratio, since there is no load resistor to contribute unwanted noise. The capacitance of the diode junction limits frequency response, so the photovoltaic mode is generally used to detect low-level dc and low-frequency signals. The photoconductive mode produces more sensitivity than the photovoltaic configuration. A diode connected in the photoconductive mode is essentially a current source, with a linearity extending over seven decades of incident light intensity. Both modes of operation are shown in Fig. 2-27.

A pin photodiode, operated in the photoconductive mode, is ideal for detecting relatively fast pulses from an LED in the presence of

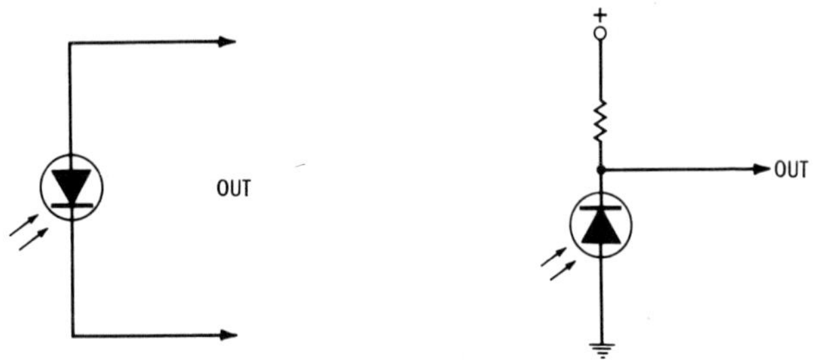

(A) Photovoltaic mode. (B) Photoconductive mode.

Fig. 2-27. Photovoltaic and photoconductive operation.

ambient light. As with the phototransistor circuits described earlier, a high load resistance gives high sensitivity while a small load resistance gives fast response times. Several circuits for amplifying the photocurrent generated by a photodiode in the photoconductive mode are presented in Chapters 5 and 6.

A photodiode can be operated into an op amp to provide both high sensitivity and fast response time. A typical circuit is shown in Fig. 2-28. Gain of the circuit can be varied by adjusting the feedback resistor R_f. Variations of this circuit are employed in radiometers and miniature detector-amplifier modules manufactured by several manufacturers of photodiode products.

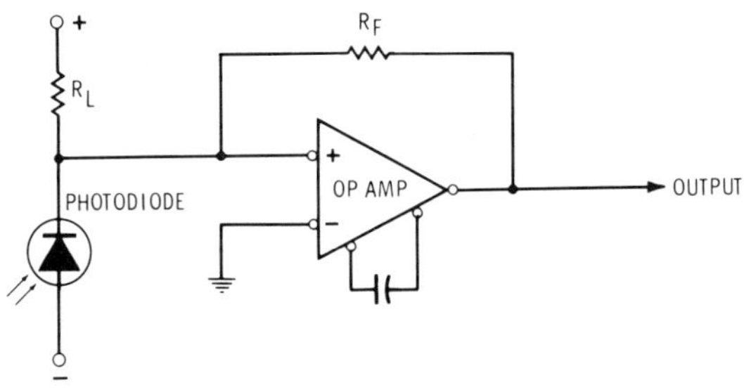

Fig. 2-28. Typical photodiode-operational amplifier circuit.

USING LEDS AS DETECTORS

In some applications, it is convenient to use an LED in a detector mode. LEDs are sensitive to the wavelength emitted by LEDs composed of identical semiconductor; so best results are had by matching an emitter LED with a detector LED of the same type number.

LED detectors can be operated in both the photovoltaic and photoconductive mode. Amplifier circuitry, similar to that presented for the solar cell and photodiode on previous pages, can be used with an LED detector with good results.

An LED operated in the detector mode permits several interesting and unique applications. An optical communications link employing a single LED as both emitter and detector is a particularly attractive application, since only one optical element is needed. In the case of an atmospheric optical link, only a single lens is required, while a fiber-optic link requires only a single strand. The space savings and economy of this approach are superior to systems employing a separate emitter and detector. Both the lens and fiber-optic approaches permit two way communications over the same optical path.

LEDs as detectors can also be employed in two-way optical repeaters. Like conventional telecommunication links employing wires, fiber optic links require repeaters to boost the transmitted signal at appropriate intervals. Application of LEDs in two-way repeaters and source-sensor pairs is discussed in Chapter 3.

CHAPTER 3

Source/Sensor Pairs

Several circuits in Chapter 2 depend on the compatibility of an LED source and a sensor. There is an entire family of devices which exploit the various possibilities of this relationship and we will discuss them under the general name *source/sensor pairs.*

Soon after the first commercial LED was introduced in 1962, several optoelectronic devices containing an integral LED and photodetector became available. The simplest contained a GaAs hemispherical LED and a silicon phototransistor and was called a photon-coupled transistor. A more sophisticated device was the optoelectronic pulse amplifier. This device contained an integral LED, silicon detector, and amplifier in a miniature integrated circuit flat pack. A cross-section of one of these units is shown in Fig. 3-1. By providing total electrical isolation between an incoming and outgoing signal, the unit made possible noise-free interfaces in computer subsystems and data transmission systems. Several more advanced versions of these optoelectronic couplers are being marketed today.

Electro-optical isolation circuits comprise the biggest application for source/sensor pairs. There are several other applications, however, and several will be reviewed in this chapter. First, opto-isolators will be described in more detail.

OPTO-ISOLATORS

One of the biggest applications for LEDs in electronic circuits is electrical isolation. By mounting an LED near a light sensor, a great variety of control functions can be achieved without the need for direct electrical contact. The result is that many forms of control and switching can be achieved with almost infinite electrical isolation. Opto-isolators are also called photon couplers, optoelectronic cou-

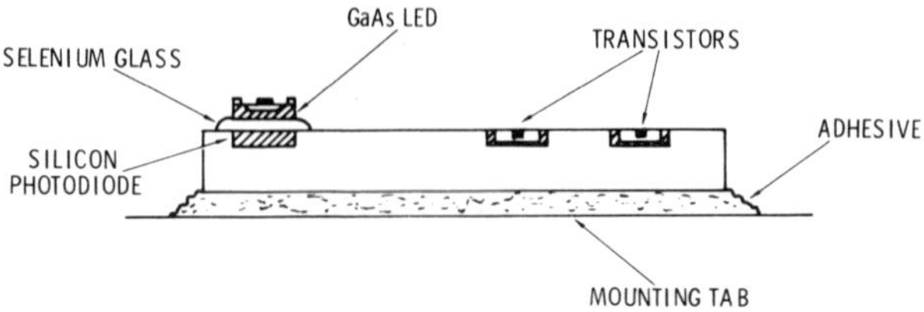

Fig. 3-1. Texas Instruments' optoelectronic integrated pulse amplifier.

plers, light-coupled circuits, coupled pairs, optically coupled isolators, and other names, but they all refer to devices that contain at least one optically coupled LED and sensor.

The most common opto-isolators employ either a visible or infrared emitter, and photoresistive or silicon detector. Photoresistive detectors are usually cadmium sulfide (CdS) or cadmium selenide (CdSe), while silicon detectors include phototransistors, light-activated SCRs, photo FETs, and pin diodes. Fig. 3-2 shows the construction of a typical commercial isolator.

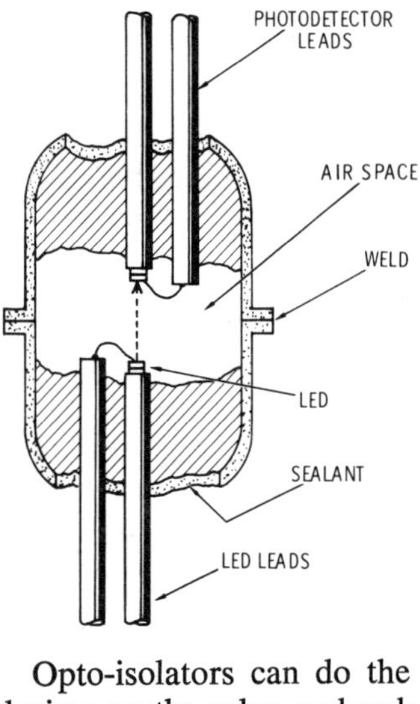

Fig. 3-2. Typical opto-isolator construction.

Opto-isolators can do the job of such traditional electromagnetic devices as the relay and pulse transformer and do it with solid-state reliability, speed, and size. Other advantages include low cost, electrical isolation, absence of contact bounce, and low power requirement. The chief drawbacks of solid-state isolators include finite *on* and *off* resistance, limited current capability, and low transfer efficiency.

Numerous kinds of opto-isolators are available commercially, but it is possible to assemble custom units with a minimum of difficulty. For example, if a nearly infinite isolation impedance is required, an

LED and photodetector can be connected by means of an inexpensive length of optical fiber. For more conventional applications, all that is needed is to cement or tape the two components together to provide good optical coupling.

OPTO LATCHING RELAY

An LED-SCR opto-isolator can be used as a simple latching relay. As shown in Fig. 3-3, the LED activates the SCR gate and turns it *on*. The SCR stays *on* in a latching mode until the reset switch is momentarily opened.

This circuit can be used to control small motors and other electromagnetic devices. The advantage of the circuit is that total electrical isolation is achieved via photon coupling. In sensitive digital equipment or amplifier circuits, isolation is frequently necessary to prevent erroneous signals and noise. The isolator also prevents high-voltage spikes, sometimes generated by inductive circuits, from being transmitted back into the control circuitry.

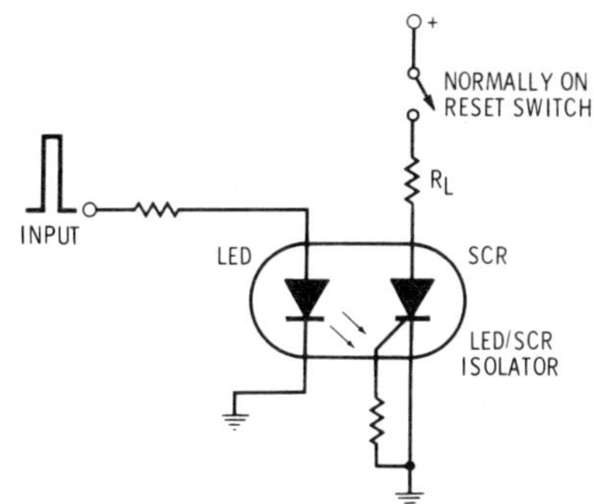

Fig. 3-3. Opto latching relay.

PULSE INVERTER

The SCR opto-isolator just described can also be used to provide both a positive and negative pulse output in response to an incoming LED pulse. A suggested circuit is shown in Fig. 3-4. In operation, capacitor C1 charges through R1 until the SCR is activated by the LED. Capacitor C1 then discharges through the SCR and load resistor R3 to give a negative output pulse. Simultaneously, a positive output pulse appears at the anode of the SCR and load resistor R2.

LOGIC AMPLIFIER

A phototransistor opto-isolator can be used to couple a logic level from one point to another. A typical circuit diagram is shown in

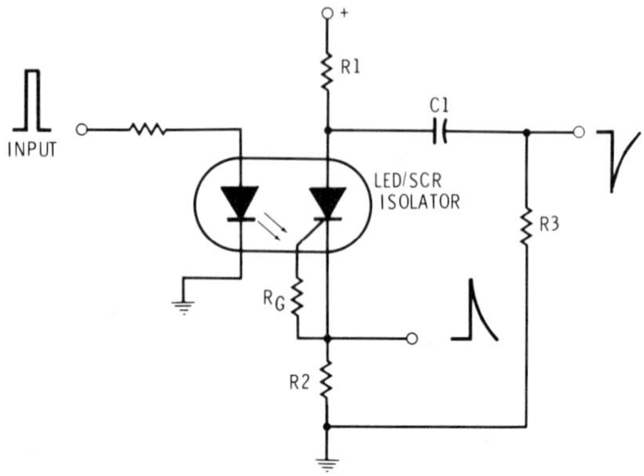

Fig. 3-4. Pulse inverter.

Fig. 3-5. Normally, Q1 is *off* and no current flows through the LED, but a pulse at the base of Q1 turns the transistor *on* and the LED is forward biased. With the LED *on,* the opto-isolator phototransistor is also turned *on,* and the pulse is transferred optically.

LOGIC INVERTER

The circuit shown in Fig. 3-6 inverts an incoming logic signal while ensuring complete electrical isolation with a source/sensor pair. In operation, a high logic level at the LED activates the phototransistor and results in a high level at the input of the inverter. This signal is then inverted and appears as a low logic level at the output. The reverse occurs when a low logic level is transmitted through the isolator. With no LED forward current, the phototransistor does not conduct, and the inverter output displays a high logic level.

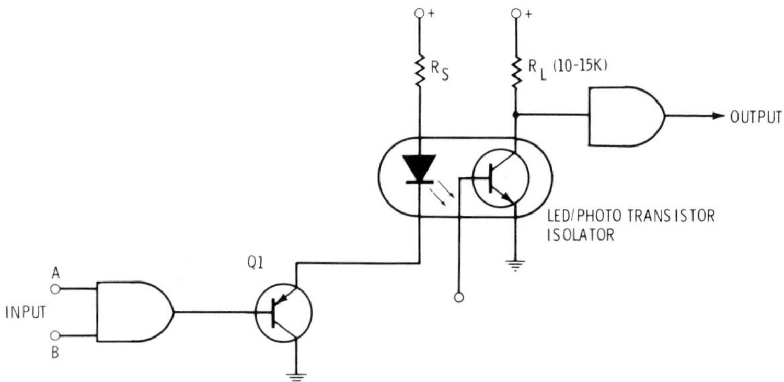

Fig. 3-5. Isolated logic amplifier.

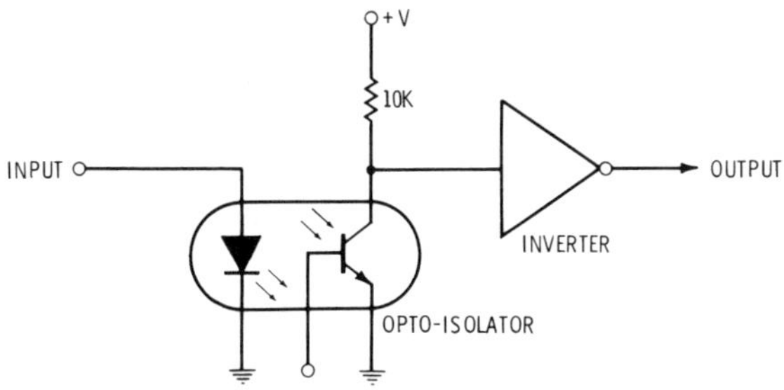

Fig. 3-6. Optoelectronic logic inverter.

OTHER CIRCUITS

Manufacturers of opto-isolators publish application notes and brochures providing additional circuits for source/sensor isolation circuits. Litronix, Monsanto, Motorola, Texas Instruments, and other manufacturers can be contacted directly for additional information.

SPEEDING UP OPTO-ISOLATORS

In some applications, it is necessary to employ an opto-isolator capable of receiving and transmitting signals with a frequency in excess of one megahertz. Frequencies of this magnitude are easily handled by GaAs and GaAsP LEDs, but some detectors, particularly phototransistors, do not normally operate at such high frequencies. The phototransistor speed-up circuits described in Chapter 2 can be easily employed to remedy this situation. By means of an appropriate speed-up circuit, frequencies of up to 10 MHz can be handled by a conventional LED-phototransistor opto-isolator.

In critical scientific applications, the response time of the LED and even the speed of light itself can be limiting factors. Remember that light travels approximately one foot each nanosecond. In precise scientific applications, this transit time must be taken into account. This is particularly true when fiber optics are used to couple a source with a sensor.

CIRCUITS FOR HIGH-LEVEL LOADS

Many high-power consumption devices cannot be directly driven by a conventional opto-isolator. The isolator is just not capable of providing sufficient voltage or current. There are several simple techniques for obtaining more current or voltage from a conventional opto-isolator. Litronix, Inc., a major manufacturer of LEDs and opto-isolators, recommends the circuit shown in Fig. 3-7 for driving high-current loads with a standard phototransistor opto-isolator.

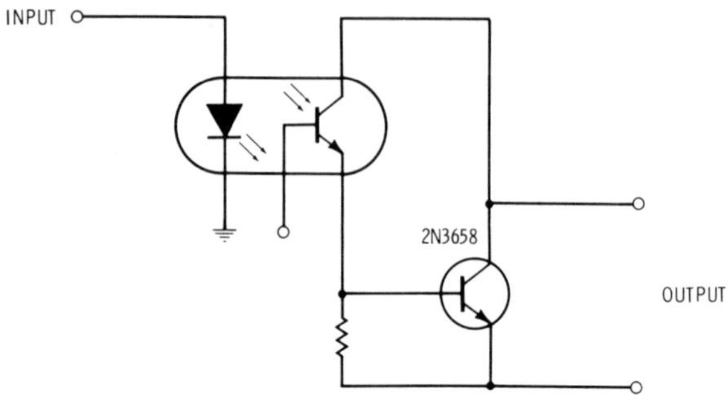

Fig. 3-7. High-level current driver.

Normally, the isolator will drive a load requiring no more than about 10 mA, but adding the single 2N3568 transistor increases the drive current by a factor of ten. Two transistors in cascade will supply even more current and permit motors and other heavy-duty electromagnetic devices to be driven directly with semiconductor circuitry.

For more load voltage, Litronix recommends the circuit shown in Fig. 3-8. The current amplification circuit shown in Fig. 3-7 is connected in a Darlington mode, but the voltage amplifier in Fig. 3-8 operates in a normally *on* mode. When the LED is *off,* the phototransistor in the opto-isolator is also *off,* and Q1 is then biased into conduction through R1. When the LED is activated, the current flow through R1 is diverted through the isolator phototransistor, turning Q1 *off*. As in the case of the current amplifiers, more transistors will increase the voltage amplification.

OPTICAL POTENTIOMETER

A contactless potentiometer can be made from a photoresistor-LED opto-isolator. In operation, the LED causes the photoresistor to alter

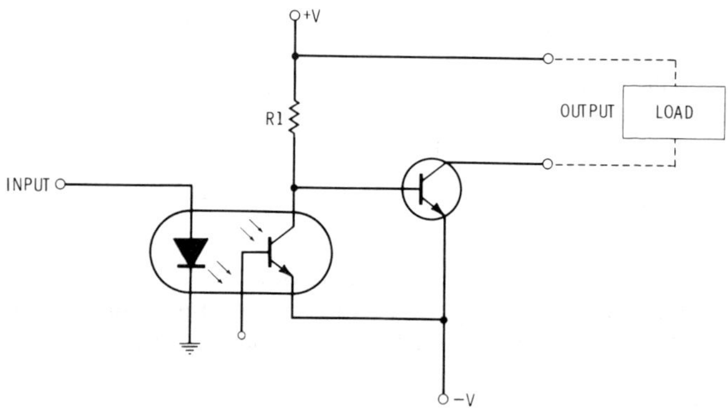

Fig. 3-8. High-level voltage driver.

its resistance in response to the intensity of the LED. As shown in Fig. 3-9, the LED is controlled by adjusting its forward current by means of a variable series resistor. The advantage of the optical potentiometer is that it provides complete electrical isolation between the control circuit and the resistive element.

Opto-isolators using photoresistors are commercially available, but a substitute can be easily made by attaching a plastic GaAsP or GaP LED to a CdSe cell with black electrical tape. The tape preserves electrical isolation and keeps stray light from the light-sensitive surface of the photoresistor.

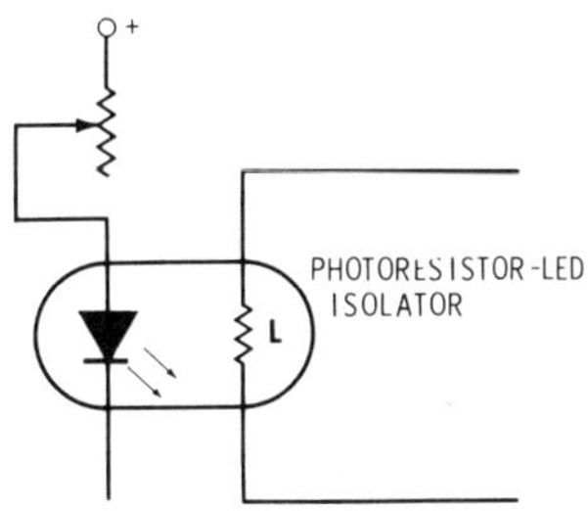

Fig. 3-9. Optical potentiometer.

OPTICAL TRANSMISSION SENSORS

A relatively new application for source/sensor pairs is optical transmission sensing. In this role, an LED and photodetector are mounted, facing one another across a slotted air gap, in a plastic package. When an opaque or partially opaque object is drawn through the slot, the LED radiation reaching the detector is reduced or eliminated, and an output signal indicating the presence of the object is produced.

A typical transmission sensor is the Monsanto MCA8 slotted optical limit switch. This component, which is shown in Fig. 3-10, consists of a GaAs infrared LED and a silicon photo-Darlington transistor. The air gap of this unit is 2.54 mm wide. The output from the detector portion of the MCA8 will directly drive TTL logic.

Applications for the MCA8 and other transmission sensors are diverse. End of tape sensing for tape recorders and cameras is one application. In the normally *on* mode, a semitransparent tape is given an opaque section at an appropriate point near its end. Opaque paint or aluminized, adhesive-backed Mylar is used for the opaque section. The tape is caused to pass through the slot in the sensor and a control signal is generated when the opaque section of tape blocks the infrared from the LED. The normally off mode is similar in operation, except an opaque tape is given a transparent section. Infrared from the LED is blocked by the tape until the transparent section passes through the slot.

A transmission sensor functions as a limit switch when employed as an end-of-tape indicator. Another limit switch application is edge sensing for sheet materials like fabric, paper, metal, newsprint, and plastic. A transmission sensor and suitable control circuitry can be employed to keep sheet materials on the track in belt sanders and in production and manufacturing processes. Mechanical limit switches used with machine tools, cams, interlocks, and foot switches can be replaced with an optical switch with greater sensitivity and much longer life. There is even a keyboard switch which employs LEDs and sensors instead of conventional mechanical techniques. As a key is pressed, an LED beam is broken, and an output signal triggered. In an alternate arrangement, the LED beam is normally blocked, and a depressed key permits a sensor to be illuminated.

Like most other source/sensor pairs, transmission sensors can play an important role in data processing. A single-element sensor, such as the MCA8, can readout holes or marks on a paper tape or IBM card with ease. Arrays of transmission sensors employing several source/sensor pairs are already in use by several computer equipment manufacturers. In the past, miniature incandescent lamps were employed for this purpose, but they have limited life, are sensitive to vibration, and are expensive to replace. An undetected burned-out bulb can cause significant and costly errors in data entry. Punched or marked LED card readers have essentially unlimited life, so there is little worry about readout error caused by a defective source. Since LEDs can be made very small, with relatively efficient and highly directional beams, there is less optical "cross talk" than in most incandescent card readers. Cross talk occurs when stray radiation from a source strikes a sensor other than its own and causes an error signal. Fig. 3-11 shows several SSL-65s, a miniature LED intended for use in computer card readers.

Finally, transmission sensors are useful in several unusual and unique applications. One is analog-to-digital conversion by means of a rotating shaft connected to a digitally encoded disc. Transparent or opaque marks on the wheel provide the output. An encoded shaft and transmission sensor can also be used to measure velocity of a liquid or gas by connecting the shaft to a turbine.

TRANSMISSION SENSOR OPERATION

Monsanto recommends the circuit shown in Fig. 3-12 for interfacing its transmission sensors to TTL logic. This circuit will permit the sensor to indicate the presence of any opaque material, but some partially transparent materials may not produce sufficient on/off contrast for reliable logic switching. Some grades of paper, for example, transmit a relatively high percentage of incident near-infrared radia-

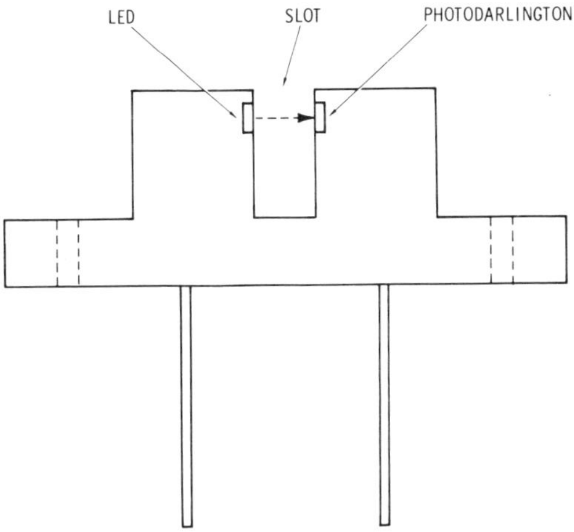

Fig. 3-10. Optoelectronic transmission sensor.

tion, and it may be necessary to install a diffusing screen in front of the LED to increase the on/off contrast ratio. An ordinary sheet of 20-lb. bond paper will increase the on/off ratio to about 5:1, but better results will be had by using a lens screen such as the Polacoat LS85PL 1/16. According to Monsanto, this screen will increase the on/off ratio to 16:1.

An easy way to measure the transmission of paper and other partially transparent materials is to use a current-regulated LED, silicon solar cell, and a 0- to 10-milliampere meter. Best results will be had with a flat-windowed LED, such as the SSL-55CF, and a large, 2-cm by 2-cm solar cell.

To perform the test, bias the LED with a fixed forward current of, say, 50 mA. Place the LED very near (but not touching) the solar

Courtesy General Electric

Fig. 3-11. SSL-65 miniature pill package infrared LEDs.

cell surface and record the current generated by the cell. A slight space should be left between the LED and cell to simuate the separation caused by the paper. Next, a square of paper larger than the solar cell is placed over the cell, and the LED is pressed against paper and cell. The new current reading is recorded and is divided by the first to find the transmittance.

This procedure was accomplished for a clear space on a page in a book. When biased at 100 mA, the LED generated a current of 2.8 mA from the exposed solar cell and a current of 0.8 mA when a section of page was placed between cell and LED. Dividing the latter by the former gives a transmission of 28.6%. Be sure to let the LED operate for a few seconds or so before performing the test, since a high forward current generates heat which reduces the optical output of the

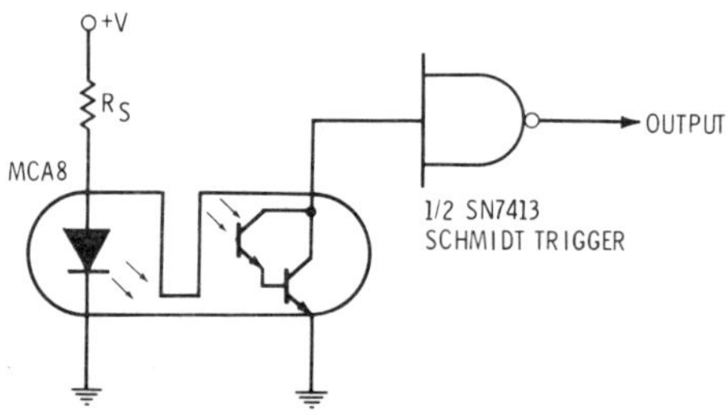

Fig. 3-12. Logic interface for MCA8 transmission sensor.

diode. Also, make the two measurements right after one another to avoid drift in the LED output (again, a temperature effect). Finally, ambient light must be kept from the solar cell to prevent measurement errors.

Several types of paper were measured using this technique. Standard 16-lb. bond paper has a transmittance (at 940 nm) of 40%. Newsprint has a transmittance of 37.5% and a 3- by 5-inch file card a transmittance of 25%.

If the transmittance of a large number of samples is to be measured, the procedure can be further simplified by using a 0- to 1-mA meter and mounting the LED in a rigid position several millimeters from the solar cell. With no sample in the gap between, adjust the current through the LED for a full-scale meter reading of 1 mA. The transmittance of objects placed in the gap can then be directly read from the meter (0.5 mA is 50%, etc.). For best results, regulate the LED current with one of the constant-brightness circuits described in Chapter 2. Also, be sure to check the full-scale reading of the meter between measurements without a sample between the LED and solar cell. If necessary, recalibrate the circuit by adjusting the LED current for a 1-mA reading.

OPTICAL REFLECTION SENSORS

The complement of the transmission sensor is the reflection sensor. In this device, an LED is mounted adjacent to a phototransistor (or other detector) so that both look out along an identical optical axis. When a reflecting material, such as a sheet of paper, comes within a centimeter or so of the source, sufficient radiation from the LED is reflected back to the sensor to trigger an output signal. A typical reflection sensor is shown in Fig. 3-13.

Like their transmission-sensing counterparts, reflection sensors have a variety of interesting and useful applications. For example, several limit-switch configurations are possible. Unlike mechanical limit switches, optical reflection sensors sense the presence of an object remotely. This can be an important capability when it is necessary to automatically sense and stop the movement of a fragile material or matter too resilient to actuate a mechanical switch.

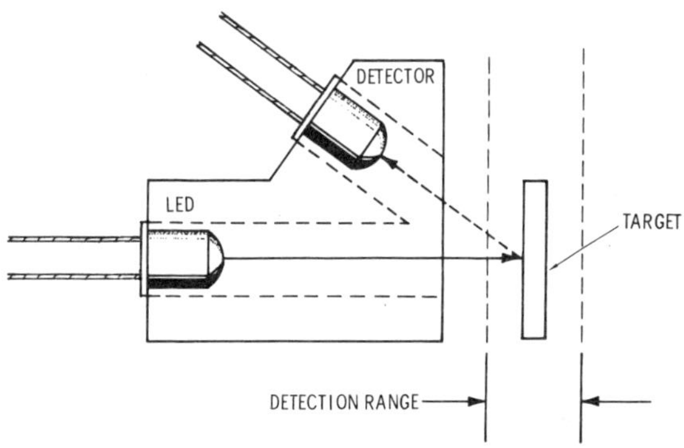

Fig. 3-13. Optoelectronic reflection sensor.

Reflection sensors can also be employed as end-of-tape indicators and in counting moving objects. A reflection sensor designed specifically to detect marked information on data processing cards is available. Designated the GP-500 Mark Sensor, this sensitive device can detect a pencil line only 0.3-mm wide when the sensor is a millimeter or more from the marking.

Reflection sensors can also be used to detect mail in a mail box, papers in a receiving location, boxes on a conveyor line, and film in a camera. Reflection sensors are ideal for use in revolution counters. The source/sensor pair is mounted adjacent to a rotating shaft, and a white dab of paint on the shaft provides a signal for each revolution.

Several novel applications for reflection sensors have been derived, and they include the detection of slight movements and vibrations. In this role, the sensor is placed adjacent to the object under test, and the detector output is monitored. Since the output signal is a function of the square of the distance to the object being sensed, very small movements will cause correspondingly large changes in the output signal.

REFLECTION SENSOR OPERATION

Reflection sensors can be operated in an analog or digital mode. If a phototransistor sensor is employed, an analog output is of questionable value due to the nonlinear transistor response to incident radiation. For this reason, thresholding circuitry to give a digital operating mode is frequently employed.

Monsanto recommends the circuit shown in Fig. 3-14, for its MCA7 Reflective Object Sensor. When sufficient reflected radiation

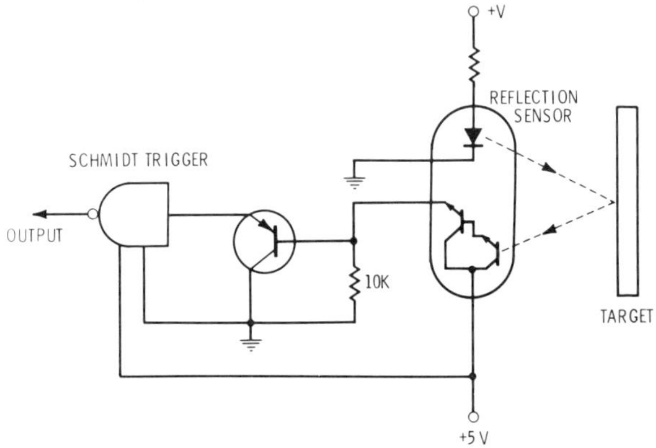

Fig. 3-14. Logic interface for MCA7 reflection sensor.

illuminates the phototransistor in the MCA7, a photocurrent is generated and passed on to a single transistor stage for amplification. The Schmidt trigger provides a threshold switching action to deliver a high-level output signal to subsequent circuitry.

For higher sensitivity, the circuit shown in Fig. 3-15 can be employed. This circuit incorporates an operational amplifier to increase low-level signals from the phototransistor. The amplified signal is then sent on to a Schmidt trigger for a threshold output action.

A particular detection application for a reflection sensor may require some experimentation. Just as transmission sensors are dependent upon the transmittance of objects being sensed, reflection sensors are dependent on reflectance. The issue is complicated by the nature of target reflectance. For example, aluminized Mylar has a very high near-infrared reflectance, but the reflectance is *specular*. Unless the plane of the shiny film is directly perpendicular to the axis of the re-

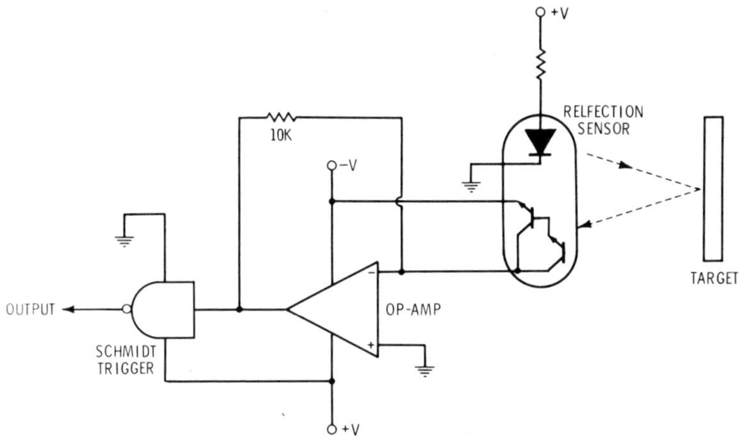

Fig. 3-15. Operational amplifier logic interface for reflection sensor.

flection sensor emitter and detector, the oncoming LED radiation will be reflected *away* from the detector. This is because a specular reflector reflects a light beam at the angle of incidence.

Diffuse reflectors, those which scatter an incident beam of light, are easier to detect. While a specular reflector offers the potential for long-range detection, alignment is very important. Diffuse objects are detected at closer ranges but with much greater reliability. Diffuse and specular reflection are summarized in Fig. 3-16.

Paper and most painted surfaces have predominantly diffuse reflectance characteristics. Many surfaces have some *gloss* or specular reflectance, but diffuse reflectance predominates and ensures reliable

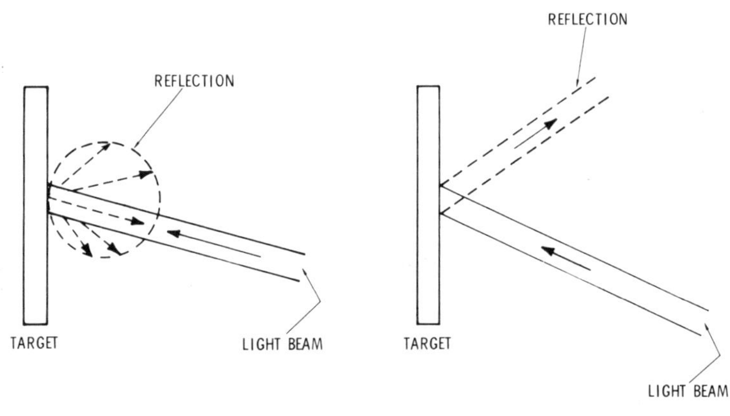

(A) Diffuse reflector.

(B) Specular reflector.

Fig. 3-16. Diffuse and specular reflectances.

detection. White paper generally has a reflectance of 70% or higher, and, as a rule, thicker paper has higher reflectance. The MCA7 supplies an output current of 0.3 mA when a white card with a reflectance of 90% is placed 1.5 cm from the sensor.

For longer detection ranges reflective tape can be used. This tape is adhesive backed and coated with a layer of tiny glass beads. The glass beads have the property of reflecting an oncoming beam of light back to its source. When used in conjunction with the MCA7 or similar reflection sensor, reflective tape permits extended detection range. The material is very thin, tough, and suitable for outdoor use.

Like transmission sensors, reflection sensors can easily be assembled from discrete LEDs and photodetectors. Care should be exercised to prevent stray light from the LED from striking the sensitive surface of the detector, since this will reduce sensitivity. Also, both LED and detector should be secured firmly in place to preclude slight movements which can affect operating sensitivity.

Finally, operation of a reflection sensor in the presence of bright, ambient light can be enhanced by pulsed operation of the LED and ac coupling the photo-detection to the output circuitry with a capacitor. Pulser circuits such as those presented in Chapter 2 will suffice for driving the LED. A capacitor in series with the detector output and the output circuitry will pass the pulsed signals from the LED while blocking the dc level from undesirable ambient sources. This technique is limited by the saturation level of the detector being used, which is low in the case of a photo-Darlington due to high internal gain, but it should improve operation in most cases.

MOVEMENT DETECTORS

Source/sensor pairs can be used in a wide variety of movement and rotation detectors, and one example is shown in Fig. 3-17. In this ap-

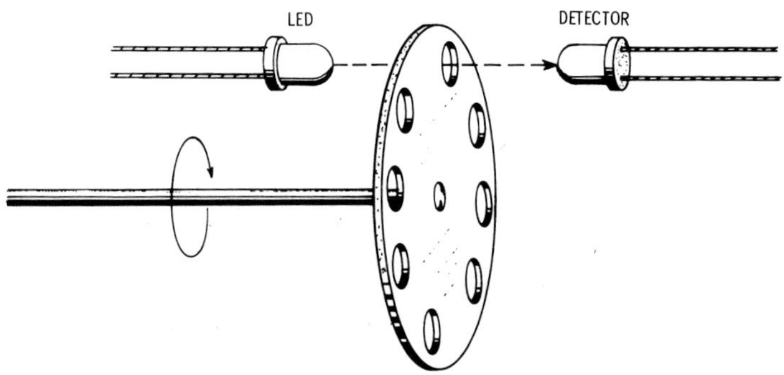

Fig. 3-17. LED rotation detector.

plication, a rotating wheel supplies a trigger pulse to a sensor each time a small hole passes by an LED source. If the wheel has only one hole, there will be one pulse for each rotation. An alternate arrangement for detecting the rotation of the wheel is to operate the source/sensor pair in a reflection mode by placing both source and sensor adjacent to one another and facing the wheel. A small square of aluminum foil or a dab of white paint will trigger the sensor on each revolution.

This general technique can be easily adapted for detecting a rotating shaft or a moving assembly. In all cases, appropriate counting logic will provide a sum of revolutions or movements. In applications where a count is not required, such as when a chopper wheel is used to break a steady beam of light into a pulsating beam, a source/sensor pair is useful as an oscilloscope trigger mechanism.

OPTICAL COMMUNICATIONS REPEATERS

Long-range optical communications links require repeater circuits to boost the transmitted signal. Similar electronic circuitry can be employed for both atmospheric and fiber-optic systems, but the optics for each system are unique.

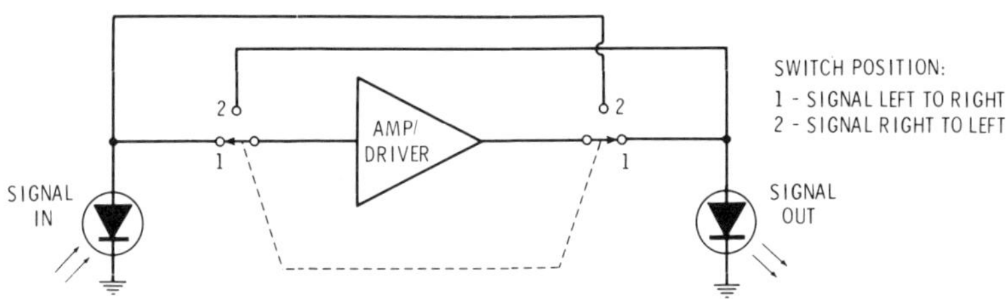

Fig. 3-18. LED-LED two-way optical repeater.

Many of the circuits described in Chapters 2, 5, and 6 can be adapted for use in optical repeaters. For two-way repeaters using the same optical path, use an LED for both source and sensor. A typical circuit is shown in Fig. 3-18. In operation, an incoming optical signal is detected by the photodiode and amplified. The amplified signal, which is a faithful reproduction of the incoming optical signal, is then used to modulate an output LED.

Optical communications with LEDs is discussed in more detail in Chapter 5. Circuits and operating hints are included.

OPTOELECTRONIC LOGIC

The discussion on opto-isolators earlier in this chapter noted the use of LEDs and photodetectors in several logic schemes. In fact, a basic opto-isolator is actually a logic device itself, since it responds to an input condition with a representative output level. The isolation characteristics of optoelectronic logic elements offer unique applications for optical logic, and several will be described here.

Optoelectronic Inverter

An LED and photoresistor can be used to make a simple optical-inverter circuit, and one possibility is shown in Fig. 3-19. With no light falling on the photoresistor, the LED is forward biased and light is produced. When light is allowed to fall on the photoresistor, however, its resistance is reduced significantly and some of the current from the LED is shunted through it. With a sufficiently high-input light level, the LED is turned almost or completely *off* as the available current is shunted through the photoresistor.

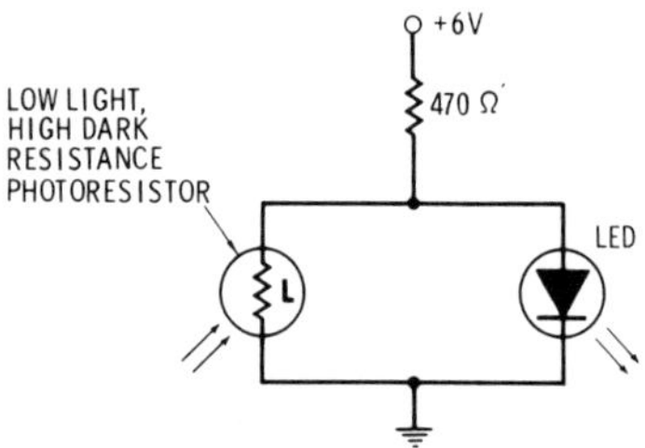

Fig. 3-19. Optical inverter.

Optoelectronic Converter

The same components used in the previous circuit can be arranged into a simple optoelectronic converter, and the result is shown in Fig. 3-20. Normally, the LED receives insufficient forward current, since the photoresistor has a high dark resistance, but light falling on the sensor lowers its resistance and forward biases the LED.

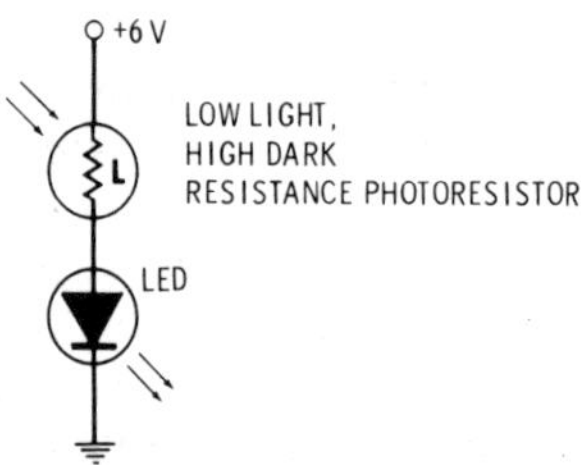

Fig. 3-20. Optical converter.

The converter circuit is unique in that it permits light with wavelengths ranging from the ultraviolet to the infrared to be converted into the LED wavelength. Conversion efficiency is not high, but the principle is of significance in understanding the potential of optoelectronic logic circuits.

With the recent advent of totally solid-state television-type image detectors, optical conversion employing two-dimensional arrays of silicon detectors (the camera) and LEDs will permit normally invisible infrared lasers, LEDs, and other infrared sources to be viewed directly. Currently, fragile vidicon or image conversion tubes are required for this application. Tubes are bulky and require considerably higher voltages than semiconductor detectors and emitters.

Memory Element

The Sharp Corporation has published an application note for its GND-50 negative-resistance LED that gives a circuit for a bistable digital memory element ("A New Opto-Electronic Functional Element," page 11). The circuit is shown in Fig. 3-21. In operation, a positive pulse triggers the memory *on,* where it remains in a conducting state. A negative pulse triggers the circuit *off* and erases the memory.

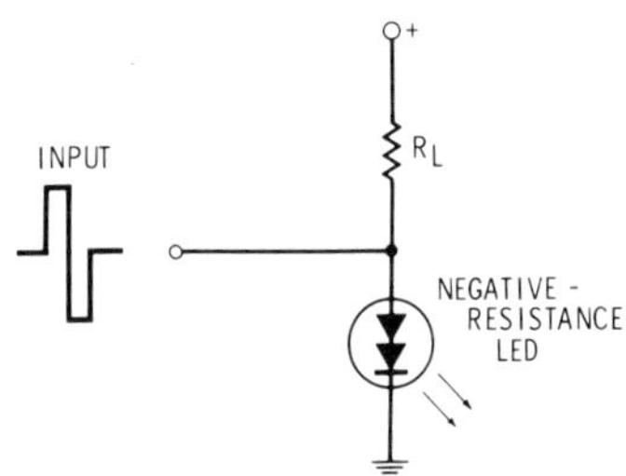

Fig. 3-21. Optical memory element.

INTEGRATED OPTOELECTRONIC LOGIC

The circuits just described are limited since they require individual components, but several fascinating optoelectronic logic circuits have been developed which use integrated circuit techniques. One example is semiconductor laser logic elements. These tiny devices perform all of the standard logic functions with tiny lasers, which switch incoming signals on command. Since light from a companion laser initiates the logic sequence, very rapid switching speeds are possible. Semiconductor lasers are described in Chapter 7.

MAKING SOURCE/SENSOR PAIRS

Many different types of source/sensor pairs are commercially available at prices ranging from a dollar upward. If a commercial source/sensor pair is not available, it is possible to make an improvised unit from an LED, a detector, and black electrical tape. The LED and detector are simply taped together in the desired configuration. An opto-isolator, for example, is made by taping the LED adjacent to the sensitive region of the detector. Sufficient tape to block external light should be employed. A reflection sensor is made by taping the LED and detector side by side so that both 'look" outward in the same general direction. For more permanent applications, the LED and detector can be epoxied together.

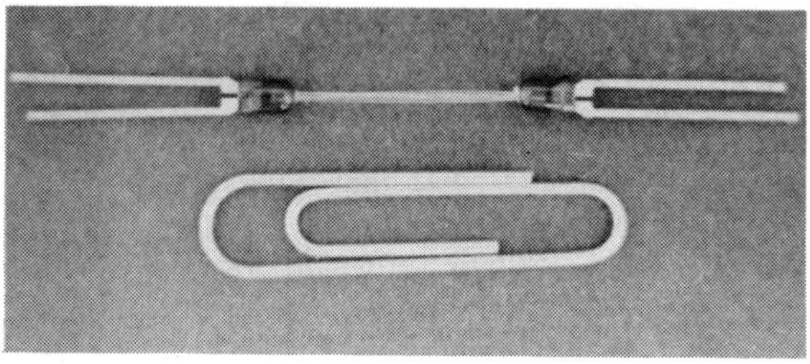

Fig. 3-22. LED-LED fiber-optic coupled, two-way source/sensor pair developed by author.

Some special-purpose circuits may require more elaborate source/sensor configurations. Fig. 3-22 shows an LED two-way source/sensor pair made by the author. This device incorporates several optical fibers to transmit a signal from one diode to the other. Since an LED is used at each end of the fiber, this optical circuit will work in both directions.

Arrays of source/sensor pairs are easily made with the help of printed circuitry. Any desired configuration can be used, so long as it can be placed on the circuit board.

CHAPTER 4

LED Indicators and Displays

By far the biggest application for visible LEDs is indicator and display devices. The advantages of solid-state operation and reliability have pushed LEDs into the forefront of modern display technology.

Numerous LED display configurations in a wide variety of packages are available from more than a dozen manufacturers. Displays include basic numeric readouts, alphabetic character arrays, linear arrays, and function signs such as "E," "+," ":," and "–." This chapter will describe each of the various displays and, when appropriate, include operating block diagrams or circuitry.

SEVEN-SEGMENT NUMERIC READOUTS

The most common display is the seven-segment numeric readout. This display incorporates seven separate rows of LEDs in the configuration shown in Fig. 4-1 to permit generation of the digits 0 through 9. Various methods are employed to make the individual segments of a display. In the *monolithic* technique, a substrate of GaAsP or other light-emitting semiconductor is masked and diffused with seven separate bars. Each bar consists of a metallized outline, usually containing several individual cells. When the metalized outline and a common electrode on the back of the semiconductor slab are activated with a forward current, the cells within the metalized outline emit recombination radiation. By controlling the bars which are activated, the various digits and several alphabetic characters can be generated.

Monolithic displays are relatively easy to manufacture since they can be made by procedures identical to those involved in fabricating very simple integrated circuits. In fact, manufacture of a monolithic display involves less steps than manufacture of a conventional bipolar IC since there is only a single diffusion step. Unfortunately, monolithic displays utilize a great deal of semiconductor material, little of which is actually involved in the light generation process. Therefore, the cost savings resulting from ease of manufacture is offset by the expense of the GaAsP substrates.

A method of making displays which employs less semiconductor material is called *hybrid* construction. In this approach, individual bars of semiconductor containing diffused junctions are soldered to a common backing. The bars are then connected to output pins and a glass or epoxy coating is applied.

The hybrid method permits large format displays to be assembled at a considerable savings in materials cost over the monolithic approach. But the cost savings is less than might be expected, since the assembly of hybrid displays is more complicated than that of monolithic units.

A variety of both monolithic and hybrid displays are offered by

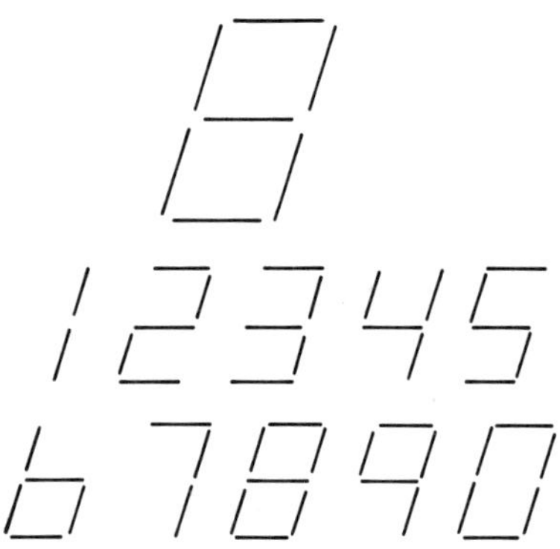

Fig. 4-1. Seven-segment numeric display.

LED manufacturers. Monolithic units are particularly popular for such high-volume applications as calculator readouts, since production costs can be made relatively low. To save material, a single digit may be made from a GaAsP substrate only several millimeters square. To improve the readability of the display, plastic bubble lenses are often encapsulated directly over the display elements. The lenses increase the apparent size of the display but limit viewing angle.

DOT MATRIX NUMERIC READOUTS

Seven-segment displays are adequate for many applications, particularly where low cost is of paramount concern. For optimum viewing, however, dot matrix numeric readouts are superior. A typical display of this type contains 20 individual LEDs in a 4 by 7 matrix. Fig. 4-2 shows the outline of one such display and the format of the digits 0 through 9.

As shown in Fig. 4-2, the digits 0 through 9 appear more legible than the angular shaped digits of a seven-segment display. This improved format is valuable in applications where a display is viewed frequently, such as on desk top electronic calculators.

DOT MATRIX ALPHANUMERIC READOUTS

Both seven-segment and 4 by 7 dot matrix readouts can generate several alphabetic characters. Generation of all alphabetic characters, however, requires a 5 by 7 dot matrix. Containing 35 individual LEDs (36 if a separate decimal point is included), a 5 by 7 dot matrix can generate at least 64 separate digits, characters, and symbols.

Fig. 4-2. Dot matrix numeric array (4 × 7).

The major drawbacks of 5 by 7 readouts are their cost and the complexity of driving circuitry. The price for a single 5 by 7 readout is usually two or three times that of a seven-segment display because of manufacturing complexity and the cost of the 35 or 36 LEDs, each of which must be matched to be within 50% brightness. Driving circuitry requires a read-only memory (ROM) to generate the various characters and symbols.

The 5 by 7 dot matrix readouts find application in advanced scientific calculators where the display is required to show both numbers and various algebraic equations, symbols, and notations. Programmable electronic calculators, such as the Hewlett-Packard 9820, incorporate a row of 5 by 7 readouts that can display words, instructions, symbols, equations, numbers, and problem results. Data entry is displayed on this advanced readout, which contains 560 individual LEDs in the case of the HP 9820, as the information is entered on a sophisticated keyboard.

Also, 5 by 7 readouts are used in keyboard verifiers. As a character is entered on a keyboard, the readout displays the entry in a convenient location and permits quick verification to catch possible mistakes.

Computer peripheral displays and avionics displays for civilian and military ships and aircraft are other applications. One unique role for the 5 by 7 dot matrix is film annotation. With separate decimal point, a 5 by 7 array contains 36 individual LEDs and provides a potential combination of an impressive 2^{36} bits. Other applications include numeric controlled machine tools, test equipment, fire control systems, and handheld radar.

LINEAR ARRAYS

Several companies manufacture special purpose linear arrays of visible LEDs which can be used in readout applications. General Electric, for example, has devised a 96-element linear display in a bar-graph format. The display includes a numeric scale and a number is indicated by illuminating all the LEDs below the appropriate digit on the scale. A variation of this scheme is to illuminate only the LED adjacent to the appropriate number on the scale. This approach requires considerably less operating power.

Monsanto markets a miniature 48-element linear array of GaAsP LEDs, the MKA1. This tiny array measures only about 6 mm long, but external optics can be employed to increase the apparent size of the array. The MKA1 is designed for film annotation, optical character recognition, and hard-copy readout. In the latter application, light-sensitive photographic paper is passed over the display while the 48 individual LEDs are activated by appropriate signals. A high-resolution chart recorder with ultra-fast response time can be made using a MKA1. Most conventional chart recorders are limited in response time since a mechanical apparatus is required to move a pen or hot wire across a moving paper. The MKA1 or other LED array permits a varying signal to be recorded on film or photographic paper as fast as the associated electronic circuitry can activate the individual LEDs. Since the LEDs have response times of about a nanosecond, no blurring will be present on the film. Other light sources have much slower response times and tend to cause blurring.

Numerous other linear arrays employing visible LEDs have been devised. Applications include readouts for depth sounders, automobile speedometers, and clock readouts. A variation of the linear display in a clock readout application is to form a circle of LEDs around a clock face. One clock made in this fashion utilized 60 separate LEDs for the minutes and seconds and 12 additional LEDs for the hours.

SYMBOL, OVERFLOW, AND POLARITY INDICATORS

Some electronic devices, particularly digital calculators and clocks, require an LED display capable of generating a special-purpose symbol, polarity sign, or colon. One such display is shown in Fig. 4-3. Many LED manufacturers offer special symbol configurations to fulfill this need. While these modified displays usually require less semiconductor material and individual LED chips, the potential savings in cost is often offset by smaller quantity purchases. For example, an electronic calculator may employ twelve or more individual digital readouts and a single polarity indicator. Since more digital readouts are required, the calculator manufacturer will receive a bigger cost break on these displays and both digital readout and simple polarity indicators may, in the end, cost approximately the same.

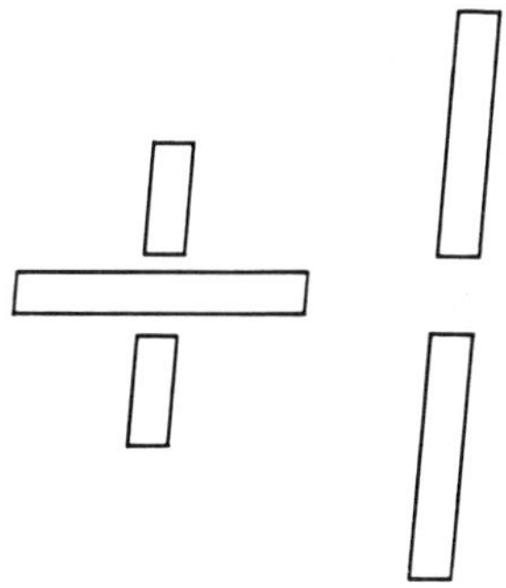

Fig. 4-3. Display ("+," "−," and "1").

The most common special purpose indicators include the "1," "+," ":," and "−." The "1" and ":" are used in digital clocks, while the "+" and "−" are used in calculators. Digital panel meters frequently employ "1," "+," and "−" indicators. In all cases the "+" and "−" signs are combined in a single package.

Another common indicator is the decimal point. Many displays include an integral decimal point, but if one is not present, a suitable alternative can be made with a single encapsulated pill package LED. This is a rare instance where a clear encapsulant is often preferrable to a diffused one, since it is desirable for a decimal point not to be a large-area light source. Discrete LEDs can also be used as error, battery status, and constant indicators on electronic calculators. In these cases a diffusing encapsulant will enhance visibility. The colon for a digital clock can be made from two individual LEDs connected in series. Again, a diffusing encapsulant will help emphasize the colon.

CHOOSING A READOUT

Whether a low-cost, seven-segment display or an expensive 4 by 7 dot matrix with integral logic is incorporated into a circuit, several specifications and characteristics should be considered before purchase. One of the most important considerations is viewing legibility.

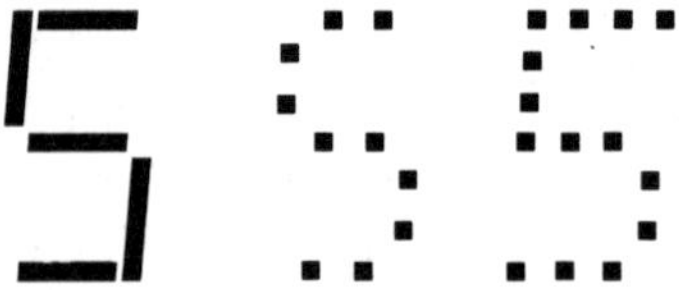

Fig. 4-4. The digit "5" as indicated on several numeric displays.

As shown in Fig. 4-4, some displays are far more readable than others. The digit "5" is shown as it appears on several popular displays in the figure. The improved resolution of the dot matrix readouts makes for a very pleasing appearance, but the limited resolution and angularity of the seven-segment readouts results in a less pleasing format. The seven-segment display containing only two dots per segment is particularly unpleasant when compared to other display formats. These displays are very inexpensive since they require such a small amount of semiconductor, but the space between diodes impairs legibility. Also, the point appearance of the individual diodes in this kind of display is uncomfortable to view for more than brief times.

The photographs of actual equipment employing LED displays in Figs. 4-5 and 4-6 sum up the importance of viewing ease and comfort. The readout on the digital clock in Fig. 4-5 is inexpensive and makes for an economical clock, but the two LED chips per segment format is uncomfortable to view. A somewhat more expensive display would provide a far better readout.

Fig. 4-6 shows two pieces of digital equipment which use better quality LED displays. Note how the uniformity of the illuminated segments enhances the appearance of the equipment and makes for comfortable viewing.

Contrast is another important consideration in the selection of an LED readout. Displays encapsulated in clear epoxy are difficult to read in the presence of bright, ambient light due to internal reflections from the substrate, metalized electrodes, connection wires, and the

Fig. 4-5. Digital clock with low-resolution, 4 × 7 display.

LED chips themselves. Since the power density of reflected ambient light, particularly sunlight, meets or exceeds that at the surface of the LED chips, the display will be only partially visible, if at all.

Contrast can be considerably enhanced by placing a red filter over the LED display, besides blocking most of the reflected ambient light while permitting the LED light to pass, the filter greatly reduces distraction caused by the internal structure of the readout. A similar effect can be gained by using a circularly polarized filter instead of a red filter.

Most manufacturers offer displays in both clear and red encapsulation. For applications where an external filter is inconvenient or too costly, it is generally wiser to choose a unit with integral filtering. Keep in mind, however, that an external filter can be more attractive when used over a row of displays than if each display is recognizable as a separate unit.

Character dimensions are another important display consideration. Some monolithic displays are so tiny that they cannot be comfortably viewed from more than a meter away. Some monolithic displays come with an integral bubble or cylinder lens which increases the apparent size of the readout, but lenses generally restrict viewing angle.

Display packaging is still another consideration. Individual displays are more difficult to install than single packages containing three or

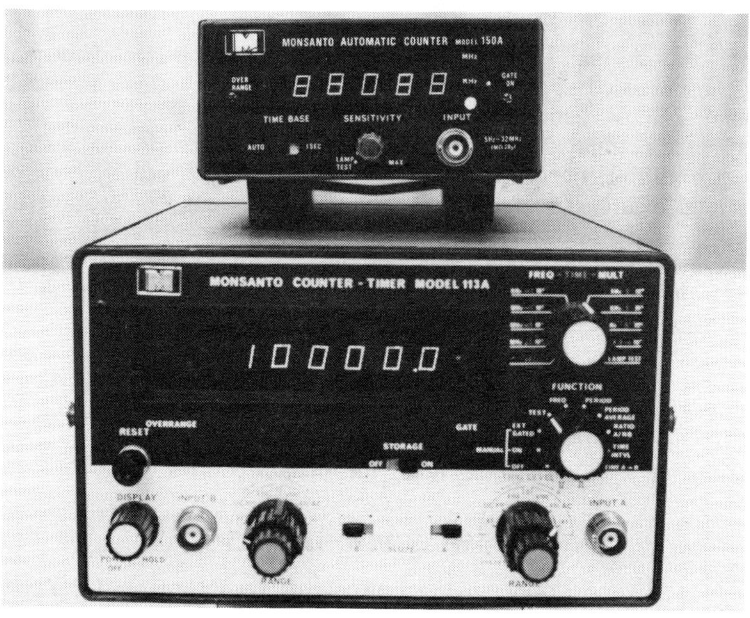

Fig. 4-6. Digital equipment with high-resolution, seven-segment display.

more readouts. Many types of multiple displays are available, and most fit a standard dual in-line IC socket. While multiple displays are easy to use and ensure even alignment from digit to digit, replacement is more expensive if only one readout in a single package becomes defective. While only one segment of one readout may be bad, in many cases the entire module will require replacement.

The final, and perhaps most important, consideration in choosing a display is cost. Invariably, displays with inferior viewing characteristics are the most economical, but some displays may have characteristics superior to those of similarly priced units. With the onset of mass production of LED displays, there is a thriving market for surplus readouts at a considerable savings over equivalent units purchased from OEM suppliers. Surplus dealers advertise their products in the pages of the various electronics experimenter magazines, so there is no need to provide names and addresses here, particularly since new firms are always entering the surplus field. Be sure to use caution when buying displays from surplus dealers. The cost savings may not be worth possible defects in the quality or construction of the device.

Summing up, it is impossible and impracticable for a manufacturer to accurately describe the viewing characteristics of a particular LED readout. Therefore, for best results, look before buying—particularly if a quantity purchase is contemplated. Sometimes a very inexpensive readout will appear so inferior that a more costly unit will become a far more attractive alternative. Conversely, a personal viewing of several readouts may reveal that an economical unit is perfectly adequate for an intended application. Potential volume purchasers can usually obtain sample readouts from manufacturers upon request. Also, some sales representatives are equipped with viewing boards containing a variety of different readouts. While this presentation usually does not include readouts made by competitors, it does enable the most appropriate and appealing devices from a particular product line to be quickly selected. Many manufacturers second source the readouts made by others, so a session with a viewing board may actually be equivalent to seeing competitor products as well.

Experimenters are usually not as concerned about absolute perfection and quality in LED displays. Nevertheless, if a particular project requires more than half a dozen readouts, the prudent experimenter will consider purchasing one display for a preliminary evaluation before spending funds for all the required units.

COMMERCIAL DISPLAYS

A great variety of commercial displays is available from LED companies in the United States, England, Japan, and other countries. In only a few years, commercial units have dropped from prices exceeding several hundred dollars to less than five dollars per digit. In volume, quantities displays can be had for less than two dollars per digit. The mass production of displays has created a large surplus market, and experimenters and hobbyists can obtain both displays and decoding logic at very economical prices from discount companies which advertise in electronics hobbyist publications, such as *Radio-Electronics* and *Popular Electronics.*

Due to the very large variety of displays offered now, and the continual influx of newer units, it is not feasible to provide a complete listing of commercial displays in a general book on LEDs. However, some of the more important manufacturers will be listed along with their representative display devices.

Fairchild

Billed as the MOD line, Fairchild LED displays include individual and multiple displays and modules. Character height is 0.125 inch. The FND 20 series includes six- or ten-character arrays in a single package with plug-in connectors. Fairchild pioneered in introducing silicon-avalanche luminescent diode arrays for film annotation about a year before GaAsP displays became commercially available.

Hewlett-Packard

A firm with an excellent reputation for its test equipment, Hewlett-Packard was the first firm to offer an LED display with integral decoding logic. The firm now offers a complete line of displays with and without internal logic, including 4 by 7 and 5 by 7 dot matrix arrays and seven-segment monolithic displays. The displays are availiable as individual characters (4 by 7 and 5 by 7 dot segment). Hewlett-Packard employs its LED displays in some of its test equipment and in its sophisticated HP-9820 programmable calculator. The calculator, which has many of the functions of a small computer, uses 16, 5 by 7 dot matrix alphanumeric arrays, with a total of 560 individual LEDs.

Fig. 4-7 shows a Hewlett-Packard multiple-element display containing integral plastic-bubble lenses. The 5-element display can be installed in a standard DIP IC socket. This display is designed for use in miniature calculators and other devices requiring a digital readout. Two or more displays can be arranged to provide more digits. Reasonably low cost is assured by using monolithic construction for the LED readouts.

A large-format LED display, made by Hewlett-Packard, is shown in Fig. 4-8. By using individual GaAsP red diodes, a 1.5-inch format is achieved. The LEDs and decoder/driver logic are installed on an etched circuit board which can be easily inserted into an edge connector.

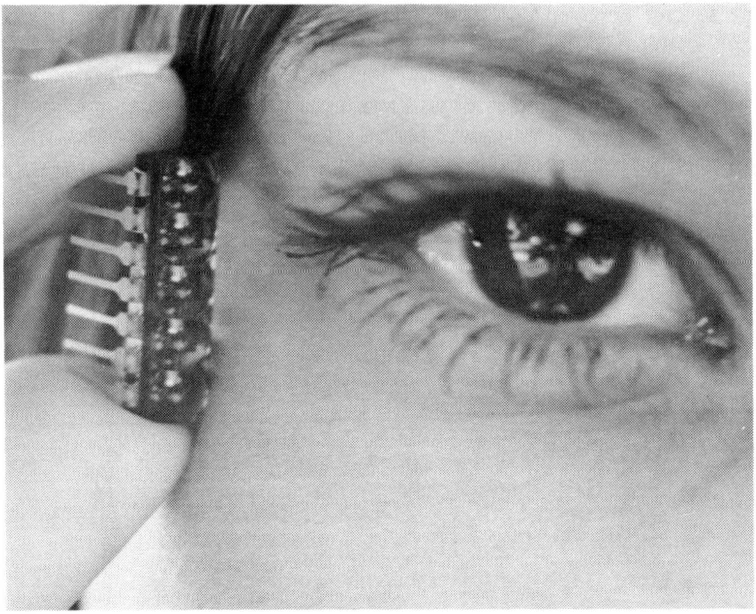

Fig. 4-7. Miniature multidigit display.

Litronix

One of the youngest firms in the LED industry, Litronix was formed in 1970. The company has experienced rapid growth as a result of an aggressive marketing program and highly competitive pric-

Courtesy Hewlett-Packard

Fig. 4-8. Display with internal logic (1½-inch-large format).

ing of its product line. Litronix manufactures a complete line of visible and infrared displays and several kinds of LED displays. At present, none of the displays incorporate integral logic, but a variety of 5 by 7 dot matrix and seven-segment displays are offered. Some of the displays are available with several characters in a single DIP package.

Monsanto

Monsanto is a pioneer in the field of LED displays. The firm offered one of the first LED displays, an array of individual GaAsP LEDs, in 1968 and currently makes a line of 5 by 7 dot matrix alphanumeric arrays and seven-segment displays, including the MAN-1. Monsanto was the first firm to market GaAsP yellow and GaP green displays.

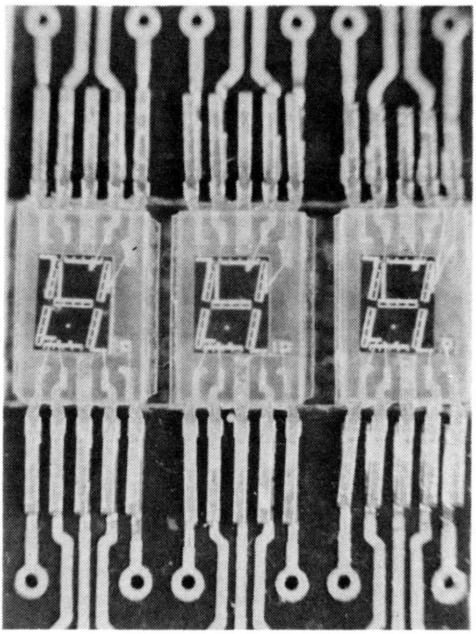

Fig. 4-9. Monolithic MAN 3 LED display.

A seven-segment monolithic display made by Monsanto is shown in Fig. 4-9. The flatpack configuration permits close unit-to-unit mounting.

Opcoa

This company offers a less diversified line of displays than any other in this condensed listing. But since Opcoa makes its displays with GaP (red), they bear mentioning. As will be noted later in this chapter, GaP displays cannot be multiplexed. However, since they are more efficient than their GaAsP counterparts, GaP displays can be a better choice in displays with separate decoding logic for each digit.

Texas Instruments

In 1962, Texas Instruments was the first firm to market a commercial LED, the infrared-emitting SNX100. Now the company complements its wide line of discrete infrared and visible emitters with a variety of seven-segment and dot matrix arrays. Both monolithic and hybrid devices are offered. Like Hewlett-Packard, Texas Instruments manufactures some displays with integral logic. Basic BCD to decimal decoding and driving logic is available, as is counting logic.

Courtesy Texas Instruments

Fig. 4-10. LED displays in hermetically sealed packages.

Texas Instruments also produces displays with a hermetic seal for sturdy industrial and military applications. Fig. 4-10 shows a 5 by 7 dot matrix and a seven-segment display made in this manner. A 6-digit monolithic array is shown in Fig. 4-11. All these displays employ a sealed glass cover to ensure a hermetic seal.

Summing Up

Product lines are by no means limited to the devices just briefly described. Many companies offer extensive varieties of character size, monolithic and discrete construction, self-contained optics, clear or red tinted potting, and various pin configurations. Several companies offer displays with either integral or modular external logic. These displays can greatly simplify eqiupment design and speed up assembly.

For more information about current product lines, pricing, and availability, contact these and other LED makers directly. The market is generally highly competitive and inquiries almost always receive prompt attention.

OPERATING DIGITAL READOUTS

Fortunately, stock integrated circuits are available for decoding a BCD signal and illuminating the proper bars of a seven-segment display. These BCD to seven-segment decoders, as they are called, can be purchased for well under two dollars in single quantities from many electronics suppliers. The TTL 7447 is the most popular seven-segment decoder IC.

Courtesy Texas Instruments

Fig. 4-11. Hermetically sealed six-digit monolithic LED display.

Fig. 4-12 shows how a typical decoder is connected to a seven-segment LED display. The transistor-resistor network provides the necessary current sinking and limiting for the decoder and LEDs respectively. Note that the decoder includes provisions for blanking (deactivating) the display and simultaneously testing all the segments for proper operation. Blanking is important in displays for counters, meters, calculators, and other instruments likely to have a lengthy

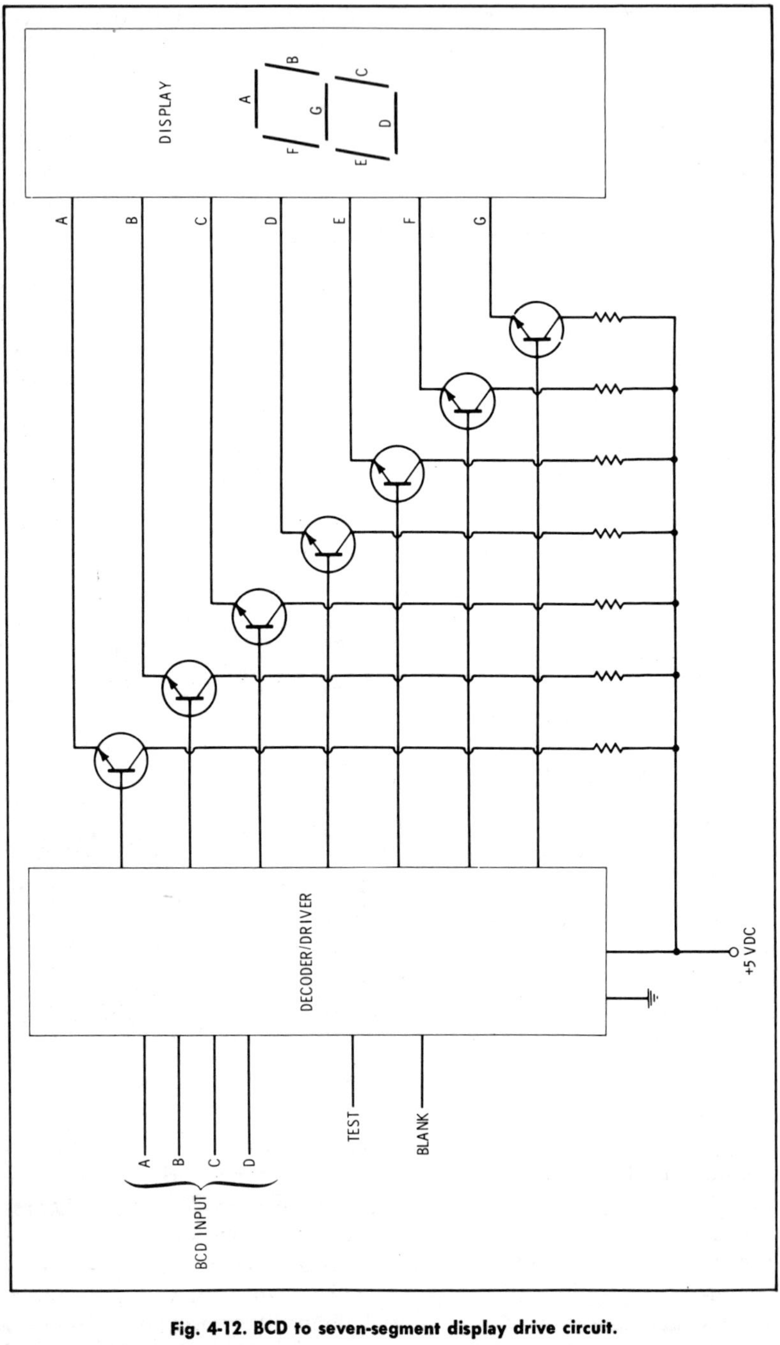

Fig. 4-12. BCD to seven-segment display drive circuit.

readout. This feature permits the extraneous zeros which would normally be present before the digits of a reading to be conveniently blanked. The test feature is equally important, since it permits a readout to be quickly checked for defective segments. An inoperative segment can cause an "8" to become an "0," or a "7" to become a "1." Some test instruments include (or should include) a display test switch which activates the test pin of the decoder IC(s). Electronic calculators usually do not include this feature, but entering a sequence of "8's" on the keyboard will provide an alternate digit check.

The transistors required between the display and decoder can be discrete devices or so-called core drivers mounted in standard IC packages. The latter approach can be very convenient, but if a single transistor should become defective the entire IC must be replaced.

When displays are operated at moderate current levels, decoders with integral driving transistors can be used. As shown in Fig. 4-13, the resulting operating circuit is considerably simplified. The current-limiting series resistors are still needed, however, since the decoder requires a 5-volt operating level.

Monsanto has gone one step further and eliminated the need for external current-limiting resistors with the MSD102 decoder/driver. This IC incorporates integral current-limiting resistors and is designed to drive the MAN 3, seven-segment display directly.

The input to these various decoder/driver circuits must be in a BCD format. Table 4-1 shows a truth table for a typical decoder/driver and gives the required BCD inputs for the digits 0 through 9, blanking, and segment test.

The operating circuitry for dot matrix arrays is more complex than that for the simple seven-segment readouts. Fig. 4-14 shows the block diagram for a commercial character generator module, the Monsanto

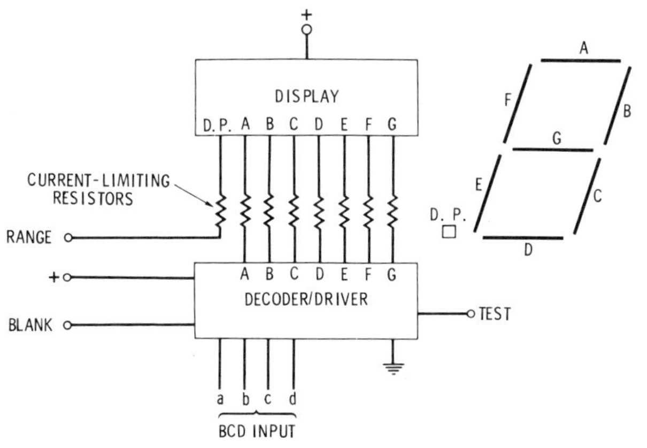

Fig. 4-13. Basic BCD seven-segment block diagram.

MDA111. This display assembly incorporates an integrated read-only memory (ROM), containing a 64-character code of 2240 bits, a clock, scanner circuitry, and drivers. A six-line BCD input activates any one of 64 distinct characters ranging from the digits 0 through 9 and the complete alphabet, to a question mark and a dollar sign.

Table 4-1. Truth Table for Seven-Segment Display

BCD Input				Output to Segments							Display
d	c	b	a	A	B	C	D	E	F	G	
0	0	0	0	0	0	0	0	0	0	1	0
0	0	0	1	1	0	0	1	1	1	1	1
0	0	1	0	0	0	1	0	0	1	0	2
0	0	1	1	0	0	0	0	1	1	0	3
0	1	0	0	1	0	0	1	1	0	0	4
0	1	0	1	0	1	0	0	1	0	0	5
0	1	1	0	1	1	0	0	0	0	0	6
0	1	1	1	0	0	0	1	1	1	1	7
1	0	0	0	0	0	0	0	0	0	0	8
1	0	0	1	0	0	0	1	1	0	0	9

DISPLAYS WITH INTEGRAL LOGIC

The ultimate LED readout would contain integral decoding logic, and in 1968, Hewlett-Packard became the first company to introduce such a device. Now a variety of displays containing logic are available from several manufacturers. Fig. 4-15 shows two Hewlett-Packard displays with integral logic.

A typical numeric display with self-contained logic is the TIL308. This display contains a seven-segment display and an internal TTL MSI (medium-scale integration) chip with latch, decoder, and driver. The MSI chip contains the equivalent of 78 individual gates.

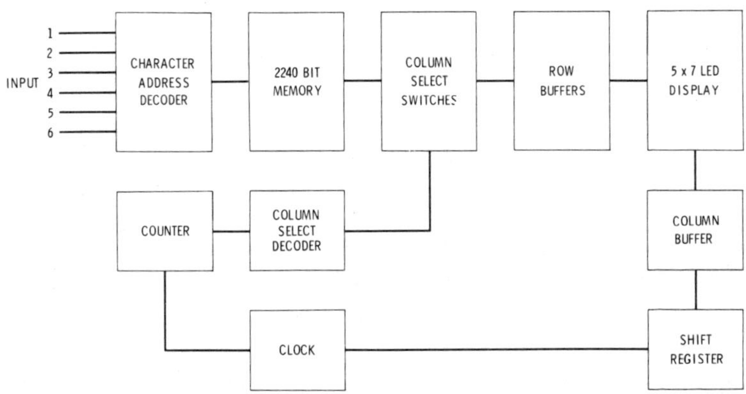

Fig. 4-14. MDA111 character generator module.

Hewlett-Packard makes several hexadecimal displays with internal logic. The 4 by 7 dot matrix provides a more stylized readout than the seven-segment approach. Texas Instruments also makes a hexadecimal display with self-contained logic.

Even more advanced are displays with internal decoding logic and counting circuitry. The Texas Instruments TIL306, for example, includes a TTL MSI chip containing the equivalent of 86 gates. The MSI chip provides a counter, latch, decoder, and driver. Provisions

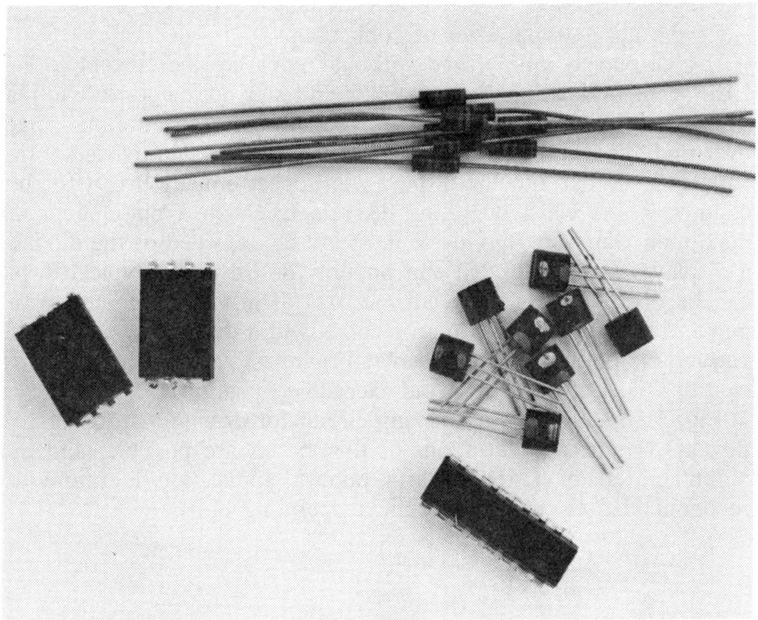

Fig. 4-15. The integral logic, in a single display on the left, replaces all the components to the right.

for interconnection permit a series of displays to be cascaded to provide a multidigit counter with a frequency capability of 18 MHz. Texas Instruments has used a combination of TIL306 counters and TIL308 displays with internal logic to construct a 100-MHz frequency counter. Suitable modifications to the basic circuit will permit an 800-MHz counting rate. The internal logic of the displays significantly reduces the component count of these counters and permits a high degree of miniaturization.

MULTIPLEXING DIGITAL DISPLAYS

It is possible to arrange a series of displays and decoder/drivers so that multidigit numbers can be displayed. When the display goes beyond several digits, however, the cost of individual decoder/drivers can become higher than the components required for a novel form of display operation called *multiplexing* or *strobing*. This technique uses one decoder/driver, which is sequentially time shared with each of the display units, one at a time. All of the respective segments of the display devices are wired in parallel, so only a connection to the anode (or cathode) of each position is required to activate a single digit in the display. A clock circuit shifts the character-enable voltage from one digit to the next, while a single BCD seven-segment decoder provides the appropriate segment connections.

If a display is multiplexed without increasing the current to each LED segment, it will appear very dim and may not be possible to read. This situation is easily remedied by increasing the strobing current over that allowed for dc operation. If the average multiplexed current equals the dc current, the display will appear equally bright for both conditions. As when operating discrete LEDs in a pulse mode, the maximum allowable duty cycle must not be exceeded or the diodes in a display will be degraded and possibly destroyed. A typical display can be operated at a peak current of 100 mA per segment, with a pulse width of 50 to 200 microseconds and a duty cycle of 20%. By varying the pulse width, the display brightness can be varied, but precautions must be taken to avoid exceeding the allowable duty cycle.

Fig. 4-16 shows a multiplexing circuit for driving a string of LED displays. Numerous variations of this circuit are possible, and most manufacturers of LED displays publish application literature with recommended circuits and detailed operating hints.

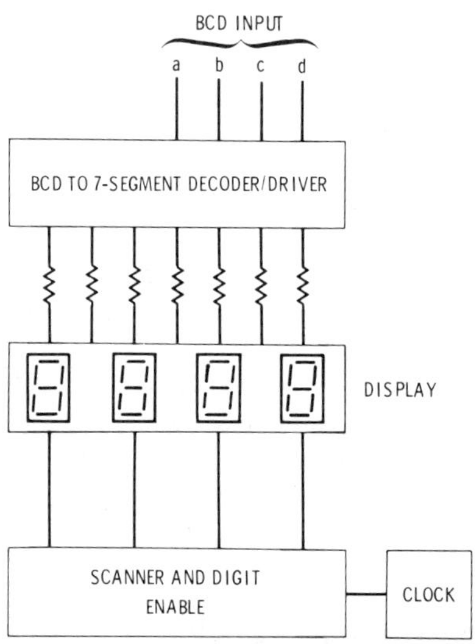

Fig. 4-16. Basic display multiplexing circuit.

The strobing technique works well with GaAsP red and yellow diodes and GaP green emitters. But, because of the inherent saturation problem of GaP red LEDs, strobing is not practical with displays made with this material. Opcoa does manufacture GaP red displays with more efficiency than GaAsP units. While they cannot be strobed, there are certain applications where separate decoder/drivers are as economical as multiplexing and where the GaP units may then be preferred as superior displays.

A few comments about the subjective viewing of multiplexed displays should be made. First, due to a slight nonlinearity in the curve of forward current versus brightness for GaAsP LEDs at very low current levels, a strobed display may actually appear brighter than a dc display with identical average current. This is because the strobing technique operates the LEDs in a more efficient mode, the linear portion of the current-versus-brightness curve. The increases in apparent brightness may also be due to the fact that the human eye is capable of responding to part of a fast pulse. Since the eye is also an integration device, in that it tends to average the light being observed, the brightness for a pulsed display may appear higher than for a dc unit operated at an identical average current.

A second point about subjective viewing of a display concerns the strobing frequency. To the eye, a light pulse rate of about 16 repetitions per second appears as a continuous glow, though this rate may vary somewhat with different subjects and viewing conditions. A multiplexed display must energize each digit at least 20 times per second to avoid any noticeable flicker from the display units. For situations where the display may be exposed to mechanical vibrations, much higher strobing rates (e.g., 100 Hz) must be employed to avoid flicker. This is because a movement of the display will cause individual digits to be energized at various points in space, resulting in a noticeable flicker.

It is particularly important to employ a fast strobing rate in digital wrist watches, pocket calculators, and other portable equipment. Under normal conditions a 100-Hz strobe rate will appear flicker free, but even the rate of 120 Hz employed in the pocket calculator of one firm flickers noticeably if the machine is rapidly moved. Of course, the device is not moved fast enough during normal operation to cause noticeable flicker.

The flicker one sees when a strobed display is rapidly moved provides a simple method of determining if the display is operated in a multiplex mode.

CHAPTER 5

LED Communication Systems

The advantages of solid-state operation make LEDs ideal for optical communicators, data links, intrusion alarms, and rangefinders. In addition to compact size, high efficiency, and low operating voltage, LEDs can be directly modulated by simply varying the bias current. With appropriate external circuitry, 100% modulation at frequencies of several megahertz or more is possible.

Transmitting and receiving systems using LEDs are dominated almost exclusively by infrared emitting GaAs and GaAs:Si diodes. Visible diodes can be modulated also, but the peak spectral wavelengths of infrared diodes closely match the peak response of silicon; thus, GaAs units are more efficient than their visible-light counterparts.

The use of infrared gives several important advantages to an optical transmitter-receiver link. Since the beam is invisible, it is very difficult to detect, a necessity for secret communications and detection-proof intrusion alarms. The invisible nature of the beam can also cause major problems for the operator by proving difficult to align without an image-converter tube or carefully aligned bore-sighting device.

Various LED transmitter-receiver systems are employed in a variety of roles, and, frequently, nearly identical systems can be used for several applications. For example, a tone-modulated transmitter-receiver pair can be used for cw communications, triangulation rangefinding, intruder detection, broken-beam counting, and other applications. In most cases, only minor circuit variations (e.g., addition of a control circuit) will be necessary to change from one application to another.

Since LED transmitter-receiver combinations can often serve in a great variety of applications, this chapter includes a general description of various modulation and detection techniques. Following this introductory material, specific systems are described in detail. Emphasis is on voice communicators, but many other applications for the systems presented will be obvious.

AMPLITUDE MODULATION (A-M)

The simplest modulation technique is to vary the amplitude of a transmitted light beam. If the amplitude variations of an input signal are directly impressed on the transmitted wave, as shown in Fig. 5-1, the input signal can be converted back to its original form at a distant receiving point. An amplitude-modulated (a-m) light source can be readily obtained by simply varying the current through an LED. Since the radiant power output will be directly proportional to the driving current when the diode is operated within rated current and temperature levels, the LED is a good source of distortion-free communication.

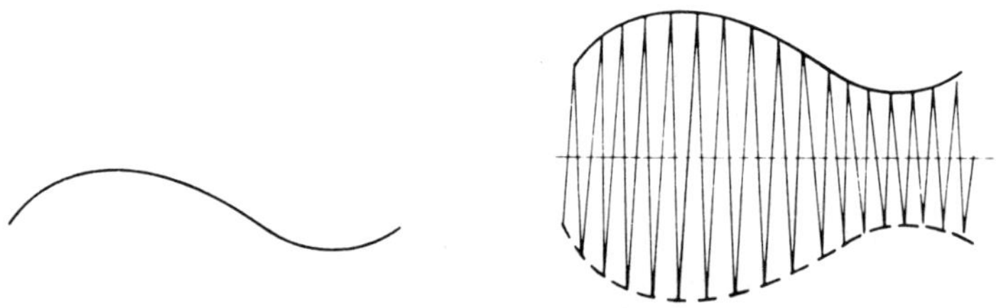

(A) Audio wave. (B) Transmitted wave (a-m).

Fig. 5-1. Amplitude modulation.

A basic circuit for amplitude-modulating an LED is shown in Fig. 5-2. In operation, the npn transistor responds to an incoming signal by varying the emitter-collector current. The LED is placed in the emitter-collector circuit and is thus directly modulated by the transistor. Resistor R in series with the LED serves to limit current to avoid exceeding the LED ratings. A voltmeter can be connected across the series resistor to determine LED current levels. With a typical incoming signal applied at the base of the transistor, adjust the resistor so that average current is a safe margin below peak allowable current. The resistor can be a potentiometer or slide-wire resistor in low-frequency applications, so long as the resistive element is rated higher than the expected power levels. Avoid wasting power at the resistor by adjusting the power-supply voltage downward as necessary.

The circuit can be used at frequencies in excess of a megahertz if an appropriate transistor is employed. Remember that GaAs:Si LEDs have a peak frequency capability of about a megahertz, and GaAs units should be employed for higher frequencies (up to 100 MHz).

The series resistor in a high-frequency modulator should be noninductive to avoid ringing, pulse stretching, and other undesirable artifacts. Also, the leads of a high-frequency circuit should be as short as possible.

Amplitude-modulated LEDs can be used to transmit voice and other signals through fiber-optic links and the atmosphere. The main disadvantage is that as the distance between the LED and detector is increased, the LED power density decreases. Depending on the optical system, the signal may be decreased as a function of the square of the separation distance. The result is that as an a-m system is operated at increased distances, the received signal experiences a rapid decrease in power density. The signal becomes difficult or impossible to detect as it approaches the background noise level.

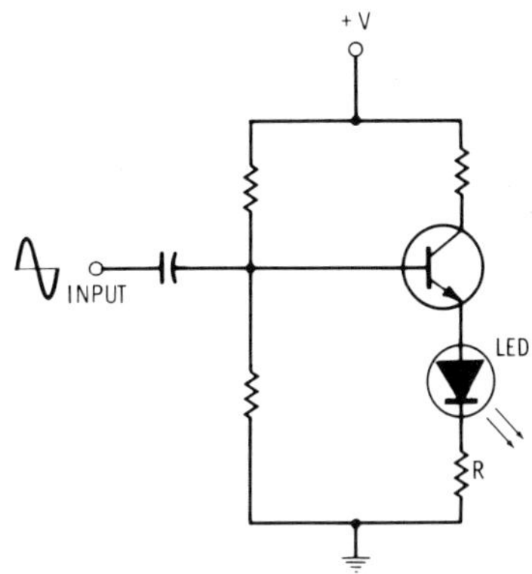

Fig. 5-2. Basic LED amplitude-modulation circuit.

In a fiber-optic a-m link, the main losses are due to the absorption within the fiber(s). Atmospheric links may be affected by a variety of constantly changing loss parameters. While the losses in a fiber-optic link may be substantial, at least they are predictable and stable. The major losses in an atmospheric link are due to water absorption, scattering from dust particles, and scintillation. Scintillation, the random distortion of a light beam as it propagates through the atmosphere, can cause significant interfering a-m modulation of its own when the separation distance between transmitter and receiver is great.

Another difficulty with a-m systems concerns the receiver. Since the receiver must be able to receive a wide range of input frequencies that may be superimposed on the transmitted beam, it is susceptible to interference from ambient light sources. An a-m receiver described later in this chapter operates well in subdued light, but it is capable of easily detecting the 120-Hz signal from a large neon sign a mile or more away.

PULSE MODULATION (PM)

The disadvantages of amplitude modulation can be largely overcome with one of several pulse-modulation techniques. This approach utilizes a train of pulses whose width, separation, rate, or combination of these parameters is modulated by an a-m or pm incoming signal. The main disadvantage of a pm optical link is the increased complexity of transmitter and receiver, but this is in many cases easily offset by the benefits of noise-free communications.

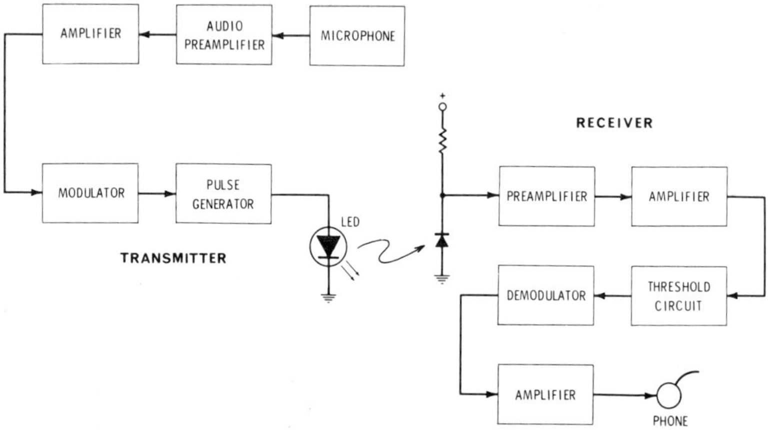

Fig. 5-3. Pulse modulation communications link.

The inherent advantages of a pm system can be seen by considering the block diagram for a simple communications link shown in Fig. 5-3. In this system, the repetition rate of output pulses from an LED is varied in direct response to an incoming signal. This type of modulation is *pulse-rate modulation* (prm), sometimes called *pulse-frequency modulation* (pfm). As shown in Fig. 5-4, the repetition rate is directly related to the input signal amplitude.

In operation, the transmitter encodes an incoming analog or amplitude-modulated signal into a variable pulse-repetition-rate light

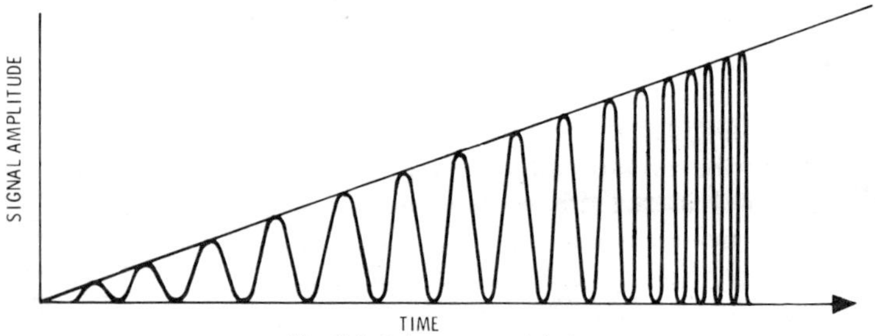

Fig. 5-4. Pulse-rate modulation.

beam. The receiver photodiode detects the pulse train and passes the signal to an amplifier circuit. The detector-amplifier circuits alone can be used to receive either a-m or prm signals, but pulse-modulated signals must be decoded by an additional circuit before they can be utilized. It is the decoding circuit that makes pulsed communication systems so attractive.

In a typical decoding circuit, the incoming train of pulses is sent from the amplifier to a logic threshold circuit. If a pulse exceeds a preset threshold value, the circuit issues a constant-amplitude output pulse. Thus, no matter what the amplitude is of a received pulse, all output pulses are equal in magnitude. Besides decoding the signal, the threshold circuit provides a noise-free detection scheme. Since the output is a train of constant-amplitude pulses, the output signal does not decrease with range. Beyond the maximum range of the system, the receiver will issue intermittent pulses or no signal at all, but within maximum range the output will be uniform and noise free.

While pulse-rate modulation is the most common, other pulse-modulation techniques are also in use. These include such schemes as *pulse-position modulation* (ppm), *pulse-width modulation* (pwm), *pulse-amplitude modulation* (pam), and others. Fig. 5-5 shows several pm schemes.

(A) Pulse-position modulation

(B) Pulse-width modulation.

(C) Pulse-amplitude modulation.

Fig. 5-5. Pulse modulation techniques.

Pulse-modulation systems are used in both communications and ranging applications. They are also useful in intrusion alarms where it is desirable to provide a false-alarm rejection for falling leaves and other natural objects that may momentarily break the beam. Since a pulsed system is on only during individual pulses, the LED can be driven at higher current levels. Therefore, a pm LED communicator can have far more peak output power—hence a greater range—than a similar a-m system. Finally, pulsed systems are ideal for use with diode lasers since most of these devices must be operated only in a pulse mode at low duty cycle.

OPTICAL COMMUNICATOR RANGE EQUATION

By means of an optical communications range equation it is possible to conveniently predict the range capability of an LED communicator. The equation incorporates such parameters as LED power and detector sensitivity and is presented below:

$$R_{max} = \sqrt{\frac{P_o A_r \tau_r \tau_a}{P_{th} \theta^2}} \qquad \text{(Eq. 5-1)}$$

where,

R_{max} is range in meters,
P_o is LED peak power in watts,
P_{th} is receiver sensitivity,
A_r is area of the receiver in square meters,
τ_r is transmission of the receiver lens,
τ_a is transmission of the atmosphere,
θ is divergence of the LED in radians.

One of the most important parameters in the equation is the receiver area, A_r. Since range is a square function, in theory, doubling the receiver lens diameter will double the range. This is because receiver area is also a square function. As an example, a miniature optical prm voice communicator designed by the author has the following parameters:

$$P_o/P_{th} = 9 \times 10^5$$
$$\theta = 0.09 \text{ radians}$$
$$\tau_r = 0.9$$
$$\tau_a = 0.9 \text{ (assumed value)}$$

With a receiver lens diameter of 12 mm, A_r is 1.13×10^{-4} square meters and Equation 5-1 gives a range of 100 meters. By doubling the lens diameter to 24 mm, almost an inch, the range is doubled to 200 meters.

The range equation may give a conservative result when compared to actual range, since the power density in a beam is frequently higher at the center. For example, an optical communicator designed by the author has a receiver area of only 0.00815 cm^2, and inserting the various parameters into the equation gives a predicted range of only 11 meters. The actual experimental range was found to be slightly more than 20 meters, almost twice the predicted value. The error should decrease as transmitter beam divergence is decreased, but for precise applications it may be necessary to derive special-case range equations. For example, if the receiver lens diameter and the desired range are known, the angle formed by the receiver alone, not by the entire beam, can be used in the equation. In this case the P_o must be the power striking the receiver, not the total beam power.

Of course, all of the parameters in the equation are important, and increasing the ratio of P_o/P_{th} and reducing the transmitter divergence will also improve range. Since divergence is also a square function, for example, halving it will double the range.

The most difficult parameter to measure is P_{th}, receiver sensitivity. A sensitive dual-trace oscilloscope can be used to obtain an approximate sensitivity level by illuminating the photodetector with a known light level while simultaneously measuring the signal level at the photodiode and receiver amplifier output. The atmospheric transmission factor (τ_a) is important only in long-range communication links. Since the spectral emission from GaAs:Si LEDs falls partly in a water absorption region, a loss factor must be included when humidity is high or when the range is more than a hundred meters. Equation 5-1 is provided as only a general guideline and is only as accurate as the values inserted into it. A variety of conditions (e.g., varying intensity across the LED beam) sometimes make accurate measurements of these values difficult.

SIMPLE A-M TONE COMMUNICATOR

A very simple a-m tone communicator can be constructed with a minimum of components. Fig. 5-6 shows the circuit for the transmitter, a simple two-transistor pulse generator and a GaAs:Si LED.

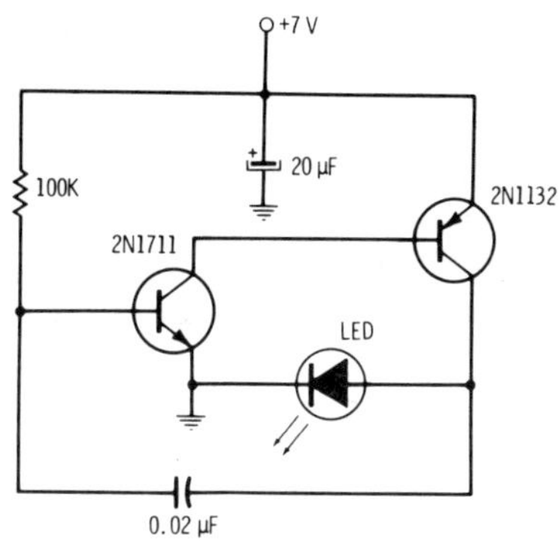

Fig. 5-6. LED tone transmitter.

The original transmitter used a domed diode, but any low-cost LED can be substituted. With the circuit values in Fig. 5-6, the LED will be pulsed with 2.5-ampere pulses at a rate of about 120 Hz. Pulse width is 20 microseconds.

Fig. 5-7 shows the tone transmitter in use. The prototype used a domed diode for wide-angle coverage, so no optics were employed. The circuit can be housed in a variety of housings, and a beam-shaping lens can be easily employed.

The circuit for the receiver used with the tone transmitter is shown in Fig. 5-8. The prototype receiver used a modular amplifier of the type found in transistor hearing aids, but any simple amplifier can be used. For more gain, use several stages of amplification. A silicon solar cell served as a detector, and no filter was used, since the receiver was designed for use in subdued light or darkness. Output was provided by a miniature magnetic earphone.

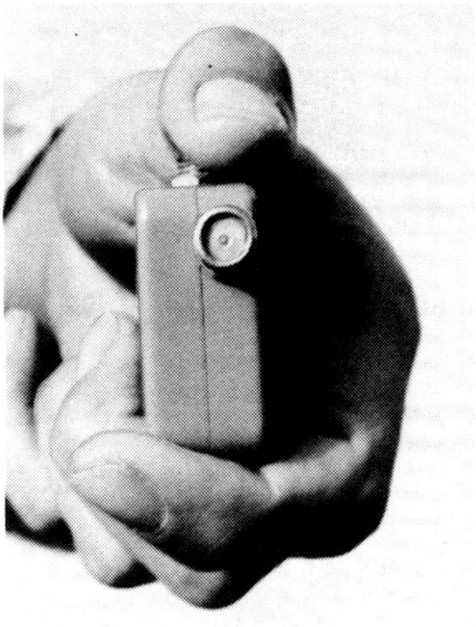

Fig. 5-7. Miniature LED tone transmitter.

The complete receiver, shown in Fig. 5-9, measures only 2.9 cm square and 1.5 cm thick (1⅛ inch square and ⅝ inch thick). A similar device can be housed in an even smaller package if space limitations make it necessary.

This simple infrared tone-communicator system was designed and constructed by the author in 1966. An experimental device, it never-

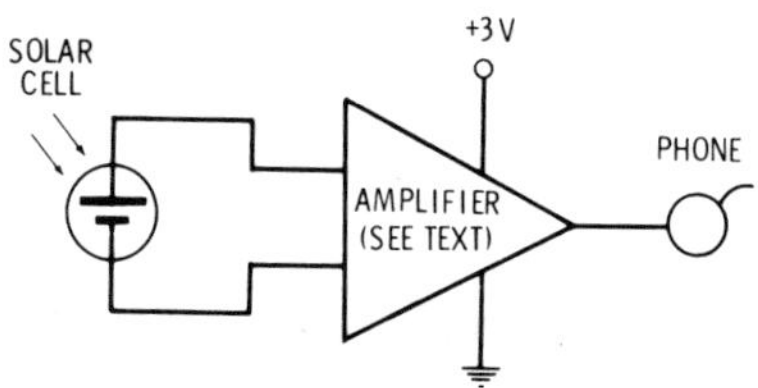

Fig. 5-8. Miniature LED receiver.

theless convincingly demonstrated the merits of an invisible-beam optical-communications link. The following circuits show how the fundamental basic concept can be expanded to incorporate voice transmission.

SIMPLE A-M VOICE COMMUNICATOR

A very simple amplitude-modulated LED voice communicator can be made with the help of commercial amplifier modules such as those

Fig. 5-9. Miniature LED receiver.

sold by electronics dealers. One such system, the "Light-Comm," was featured on the cover of *Elementary Electronics* magazine (Forrest Mims, "Light-Comm," May-June 1972). The Light-Comm article subsequently appeared in *Electronics Hobbyist* (Fall-Winter 1972) and may appear again in future magazines. The system, shown in Fig. 5-10, consists of a pair of amplifier modules, one each for a transmitter and receiver. Both systems are built into individual plastic lantern-light cases such as the Burgess "Dolphin." The system can be duplicated by referring to the original article or following the condensed instructions given here.

Transmitter

The original transmitter was assembled by mounting a modular amplifier, such as the Radio Shack 277-1240, inside a plastic lantern case. The LED in the original unit was a SSL-5C, but a newer

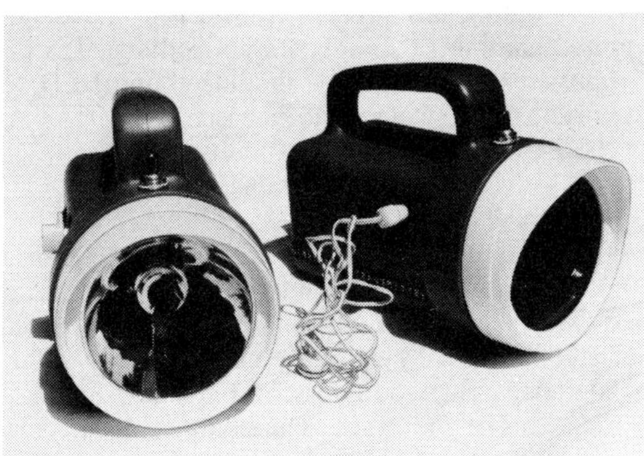

Fig. 5-10. LED a-m voice communications system (receiver on left, transmitter at right).

SSL-55C will give twice the output for about the same price and drive current. The LED is mounted to a separate board at the rear of the lantern case. A 100-ohm resistor provides current limiting for the LED. The transmitter circuit diagram is shown in Fig. 5-11.

The infrared beam of the LED is collimated by a 5-cm (2-inch) plastic lens cemented to the clear plastic reflector dust cover of the lantern light, but any simple lens with a focal length of about 11.5 cm (4.5 inches) can be used. For best results, cut a circular hole in the plastic dust cover and attach the lens to the cover with a thin bead of cement. The reflector itself is not used and is removed. Power for the

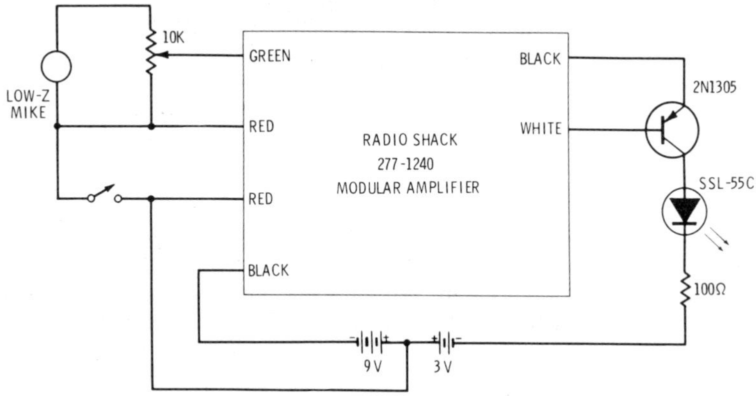

Fig. 5-11. Simple a-m voice transmitter.

amplifier is supplied by a 9-volt transistor radio battery, while the LED is operated by a pair of AA penlight cells (3 volts).

Operation of the transmitter is straightforward. The modular amplifier receives input signals (voice) from a low-impedance magnetic microphone and amplifies them. The amplified signal is passed on to a 2N1305 transistor that amplitude-modulates the LED with a drive current proportional to the signal at its base. The 100-ohm resistor in series with the LED provides current limiting.

Receiver

Assembly of the receiver is similar to that of the transmitter. The amplifier module, battery holder, and volume control are mounted to the inside of the lantern case along with a miniature magnetic speaker. Next, a detector is made by carefully mounting two solar cells back to back with transparent tape and soldering the positive lead of one to the negative lead of the other. The completed detector module is then mounted in the parabolic reflector of the receiver and held in place by its wire leads. The receiver should be wired according to the circuit diagram of Fig. 5-12.

Operating Test

Test the completed Light-Comm by pointing the transmitter at the receiver while speaking into the transmitter microphone. Voice or a strong feedback signal coming from the receiver speaker indicates the system is operational and ready for alignment. If the test is not successful, determine which device is inoperative by pointing the receiver toward an incandescent or fluorescent lamp. The speaker should issue a 60- or 120-Hz hum; if it does not, the receiver is defective. If it does, the transmitter is defective.

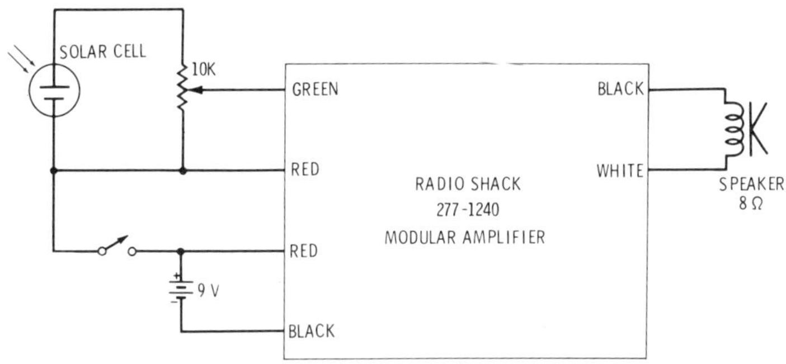

Fig. 5-12. Simple a-m voice receiver.

When both units are working properly, align the LED so that it is at the focal point of its lens, and align the solar cells of the receiver so their reflection fills the entire parabola when it is viewed straight-on at arm's length. The system can then be range tested.

Sunlight significantly impairs range capability of the Light-Comm, since there is no infrared filter over the solar cells, but at night maximum range capability will approach 300 meters (1000 feet). Both the transmitter and receiver of the Light-Comm can be used with other a-m communicators. The receiver is particularly handy for detecting signals from any of the a-m transmitters described in this chapter.

SOPHISTICATED A-M VOICE COMMUNICATOR

A more sophisticated light-beam communicator appeared in *Popular Electronics* magazine (Forrest Mims and Henry E. Roberts, "Assemble an LED Communicator—The Opticom," November 1970, pages 45–50). Designed by Henry E. Roberts, the Opticom also uses a SSL-5C LED, but its performance can be considerably enhanced by the substitution of the higher powered but electrically equivalent SSL-55C.

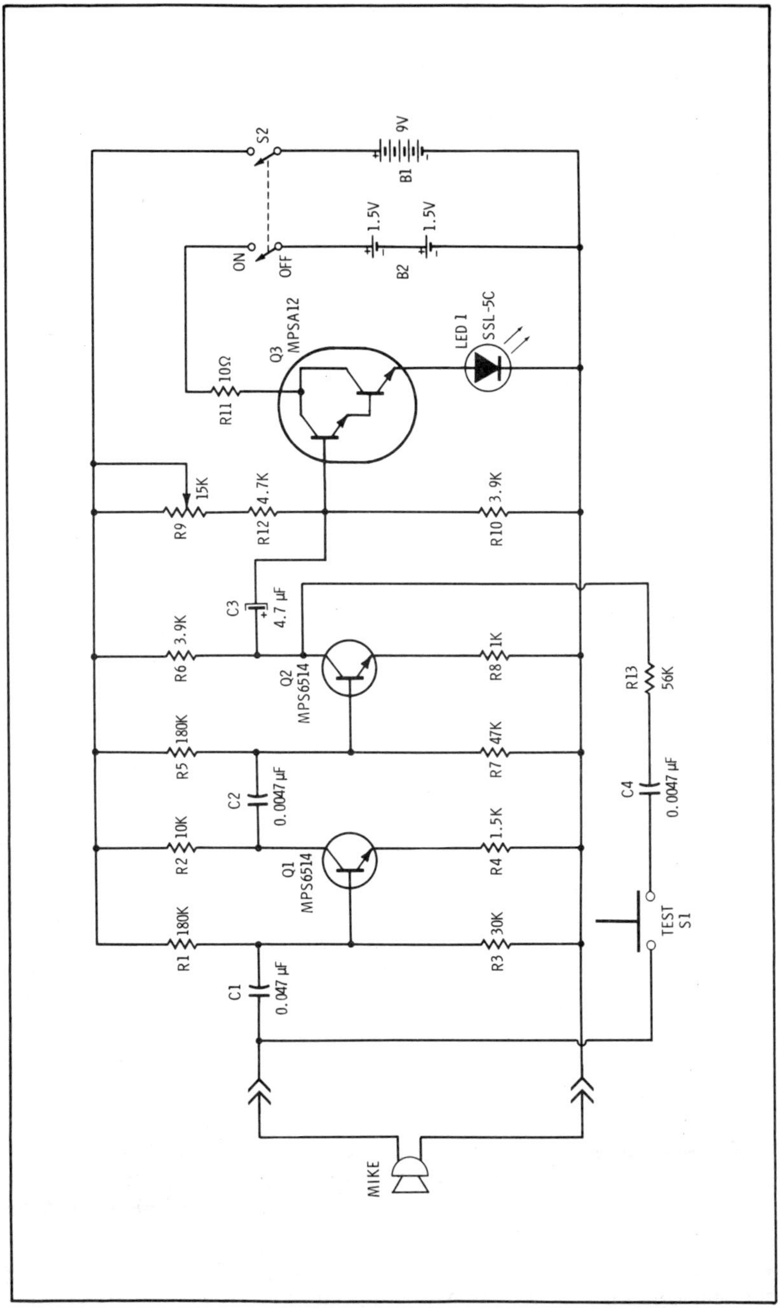

Fig. 5-13. Opticom transmitter circuit.

The Opticom transmitter incorporates a self-contained tone-generator circuit to ease optical alignment. The transmitter is simply pointed toward the receiver with the tone switch depressed until the receiver detects the signal. The tone switch is then turned off and voice communication is begun.

Transmitter

The transmitter circuit of the Opticom is shown in Fig. 5-13. The circuit can be assembled on an etched circuit board like the original version, or a perforated board can be used. In operation, a crystal microphone provides a 20-millivolt signal to the input of a MPS6514 transistor. The signal is amplified by the two MPS6514s and passed on to Darlington transistor MPSA12, which supplies bias current to

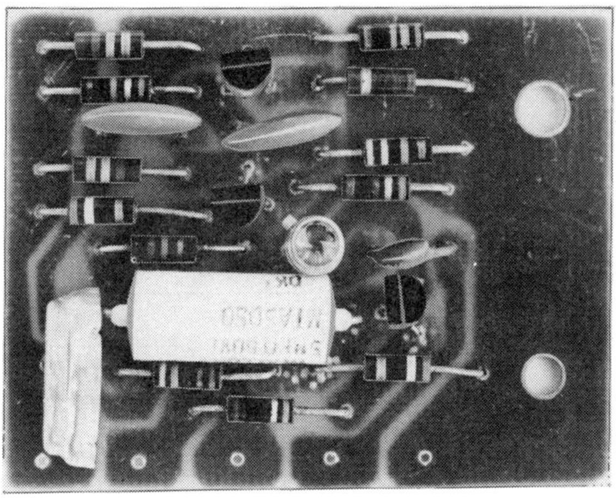

Fig. 5-14. Opticom transmitter circuit board.

the LED. (Note: If these transistor type numbers are not available, substitute good quality audio-frequency replacements, being sure to observe polarity.) The completed circuit board is shown in Fig. 5-14.

The 10,000-ohm potentiometer between B+ and the base of the Darlington transistor provides a simple way to control current through the LED. The potentiometer should be adjusted so that a 0.5-volt level appears across the 10-ohm current-limiting resistor in series with B+, the Darlington transistor, and the LED. From Ohm's law, this gives a current of 50 mA. The LED can withstand an average current of up to 100 mA without needing a heat sink, so the voltage across the 10-ohm resistor can be increased to 1.0 volt if desired. Make sure the current through the diode during operation of the transmitter does not exceed 100 mA average by observing a meter connected across the current-limiting resistor during transmitter operation.

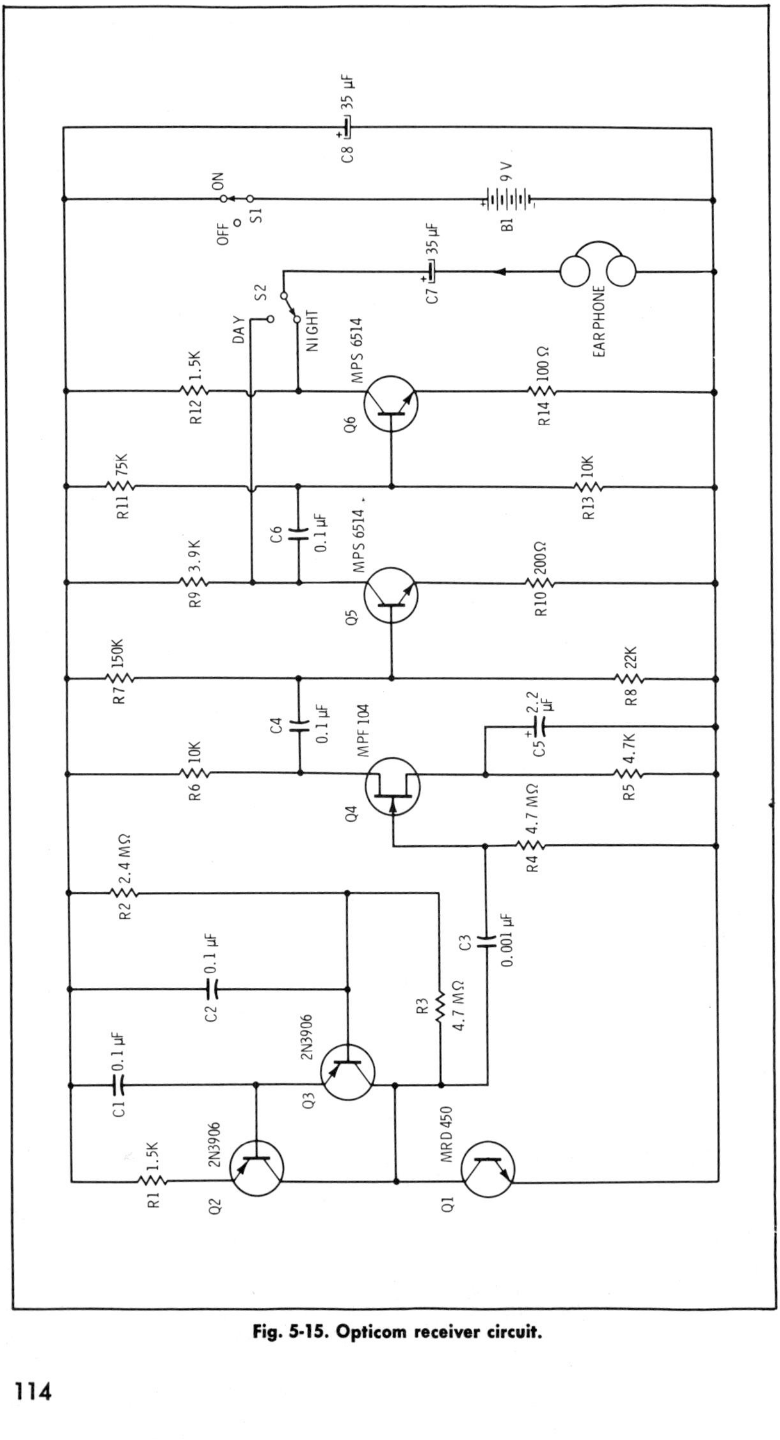

Fig. 5-15. Opticom receiver circuit.

The transmitter produces a tone for alignment or code-transmission purposes by means of a feedback circuit. When the tone switch is depressed, a 56K resistor and 0.0047-μF capacitor are placed in series between the amplifier output and input. This arrangement causes positive feedback, and a 500-Hz tone is impressed across the LED.

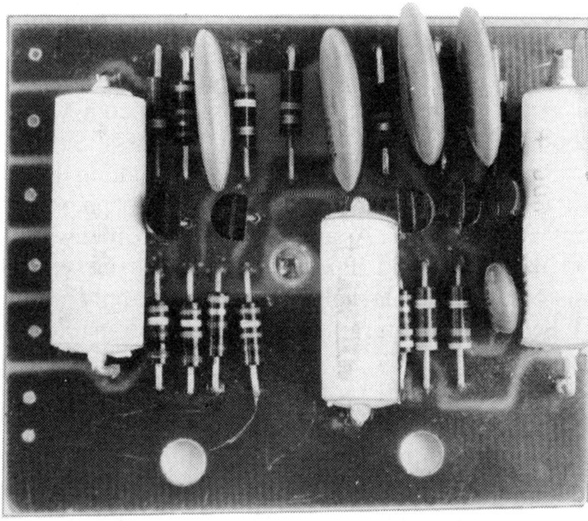

Fig. 5-16. Opticom receiver circuit board.

Receiver

The Opticom receiver uses a sensitive phototransistor as an infrared detector. The plastic-encapsulated transistor is mounted in a hole at the center of the circuit board in the original device, but numerous other mounting arrangements are possible.

Operation of the circuit can be seen by referring to Fig. 5-15. An input light signal causes the phototransistor to pass a current proportional to the signal intensity. The two 2N3906 pnp transistors form a dynamic load for the phototransistor and enhance linearity during high ambient light levels. The 2N5458 FET matches the impedance of the input circuit to an amplifier formed by two MPS6514s. For normal high-gain operation in darkness, both MPS6514s are employed, but for daylight operation, overall gain is lowered to prevent excess noise by switching out the last MPS6514. This switch should be brought out to a panel and labeled DAY-NIGHT. The completed receiver board is shown in Fig. 5-16.

Operating Test

The Opticom circuit boards should be tested before they are installed in an enclosure. This is accomplished by connecting clip leads from the boards to appropriate batteries, a microphone, and an earphone. With each board connected, point them at one another and test both the voice and cw-tone operating modes. An operating range of at least 3 to 5 meters should be obtained without the use of external lenses.

When the boards are operating properly, install them in appropriate enclosures, together with lenses to enhance transmitting range. Complete details are provided in the *Popular Electronics* article.

With 5-cm lenses at both transmitter and receiver and a transmitter divergence of 40 milliradians, the original Opticom was tested to a range of about 300 meters (1000 feet) in darkness. Substituting these values into the optical communicator range equation (Equation 5-1) and assuming a total attenuation (lens plus atmosphere) of 0.8 gives a P_o/P_{th} ratio of 9×10^4. Assuming a P_o of 3 milliwatts, this gives a P_{th} of 33 nanowatts and illustrates the excellent sensitivity of the phototransistor receiver. Daylight operation was only about 35 meters (115 feet) because of saturation of the phototransistor by ambient light. Both day and night operating range can be increased with the help of a narrower LED beam and a larger receiving lens. Good lens alignment and focusing are important for optimum results. Night operation can also be improved by biasing the phototransistor with a small GaAs or GaAsP LED. See the section on phototransistors in Chapter 2 for details. Day operation can be improved by using an infrared filter over the phototransistor and a light shield projecting outward from the receiver lens. Paint the inside of the receiver flat black for best results.

Of course, a higher-powered LED can enhance operating ranges as well. The use of an SSL-55C alone will give an output double that obtained from the less efficient SSL-5C used in the original communicator. Increasing the average current through the SSL-55C above its rated 100 mA is permissible, so long as a heat sink is employed to keep the diode at or near room temperature. More on high-power a-m transmitters follows.

HIGH-POWER A-M TRANSMITTERS

The Opticom receiver can be used with appropriate optics and a high-power transmitter to give communication ranges of a kilometer (3280 feet) or more. Two such a-m transmitters were described by Edward L. Bonin in *Electronics* magazine ("Drivers for Optical Diodes," August 10, 1964), and one of them has been modified by Texas Instruments into the circuit shown in Fig. 5-17. The circuit appears in the company's *Optoelectronics Data Book for Design Engineers,* 1972 (page 173). The device is a straightforward linear amplifier supplying a drive current of up to 1 ampere to the LED. The

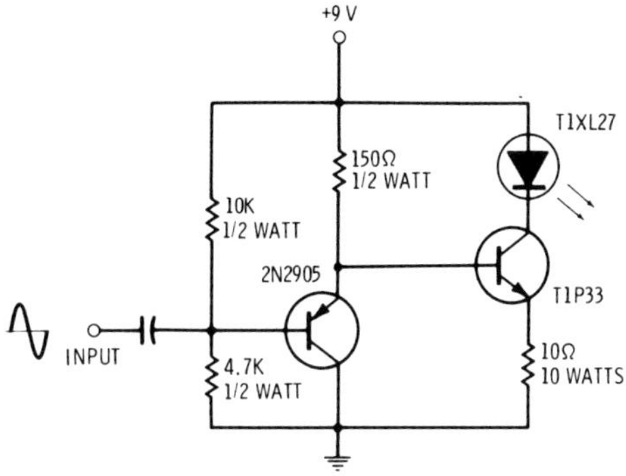

Fig. 5-17. Two-transistor LED modulator.

LED current is adjusted by varying the 10-ohm series resistor as necessary. The circuit provides 90% modulation, with a bandwidth extending from 30 Hz to 250 kHz and is ideal for voice communications. The driving transistor (T1P33) can be replaced by a high-frequency unit for operation in the megahertz range.

An improved version of the basic two-transistor modulator, also designed by Bonin, is shown in Fig. 5-18. The circuit gives 100% modulation of the LED at 1-ampere bias with only a 0.35-volt rms signal at the input. Operating bandwidth extends from 30 Hz to 25 kHz, but frequency response can be extended to much higher ranges by employing high-frequency transistors.

Both of these high-current modulators were originally designed for high average-current operation of high-powered domed emitters, but

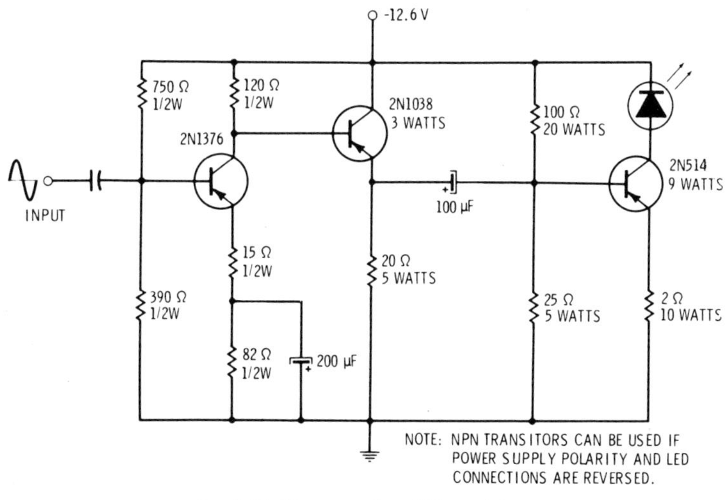

Fig. 5-18. Three-transistor LED modulator.

they can easily be modified for low-current operation. All that is necessary is to decrease the operating voltage while increasing the value of the current-limiting resistor. Use Ohm's law to monitor current with a dummy 1-ohm resistor in place of an actual LED until a desired combination of operating voltage and series resistance is found. For most efficient operation, keep the value of the series resistor low. This reduces wasted current and permits lower operating voltage. With lower operating current, inexpensive transistors can be substituted into the circuit as long as their ratings exceed expected current levels.

If operation of a conventional flat emitter at higher-than-normal current levels is desired for maximum power output, use a good heat sink. See Chapter 2 for more information on high-current operation, and remember that both high-current circuits can be used with a-m infrared receivers such as those described on previous pages.

PULSE-MODULATED COMMUNICATION SYSTEM

Full realization of the power-output potential of an LED can be realized only in a pm system. By driving the SSL-55C at 3 amperes, for example, 15 times the light power at 100-mA drive current is obtained. While pm circuits are more complex, the increased operating range justifies the use of extra components.

The Miniature Lamp Department of the General Electric Company, a major manufacturer of GaAs:Si LEDs, has designed a complete prm transmitter and receiver system. The circuits for the system and descriptive information appear in the company's 1970 booklet *Solid State Lamps—Part II, Applications Manual* (available from Miniature Lamp Department, General Electric Company, Nela Park, Cleveland, Ohio 44112). The circuits for the system are reprinted here with General Electric's permission.

The transmitter is shown in Fig. 5-19(A). An audio-modulation signal at the input is converted into a variable frequency by the 2N525 transistor. A pnp silicon transistor can be employed for better temperature stability. The pulse-generator circuit generates 0.8-microsecond pulses with an amplitude of 4 amperes at a maximum repetition rate of 50 kHz. The carrier frequency of the transmitter is determined by the adjustment of a 5K potentiometer at the audio input point.

The receiver is shown in Fig. 5-19(B). An incoming infrared signal causes the L14A phototransistor to pass a current proportional to the amplitude of the input signal. The SSL-5 LED is adjusted with the 25K potentiometer to bias the L14A with low-level infrared illumination. The 2N708 transistor provides a single stage of amplification before the detected signal is sent on to the single-shot multivibrator formed by the two 2N696 transistors. The demodulated signal appearing at the output can be converted to an audible sound with a low-cost

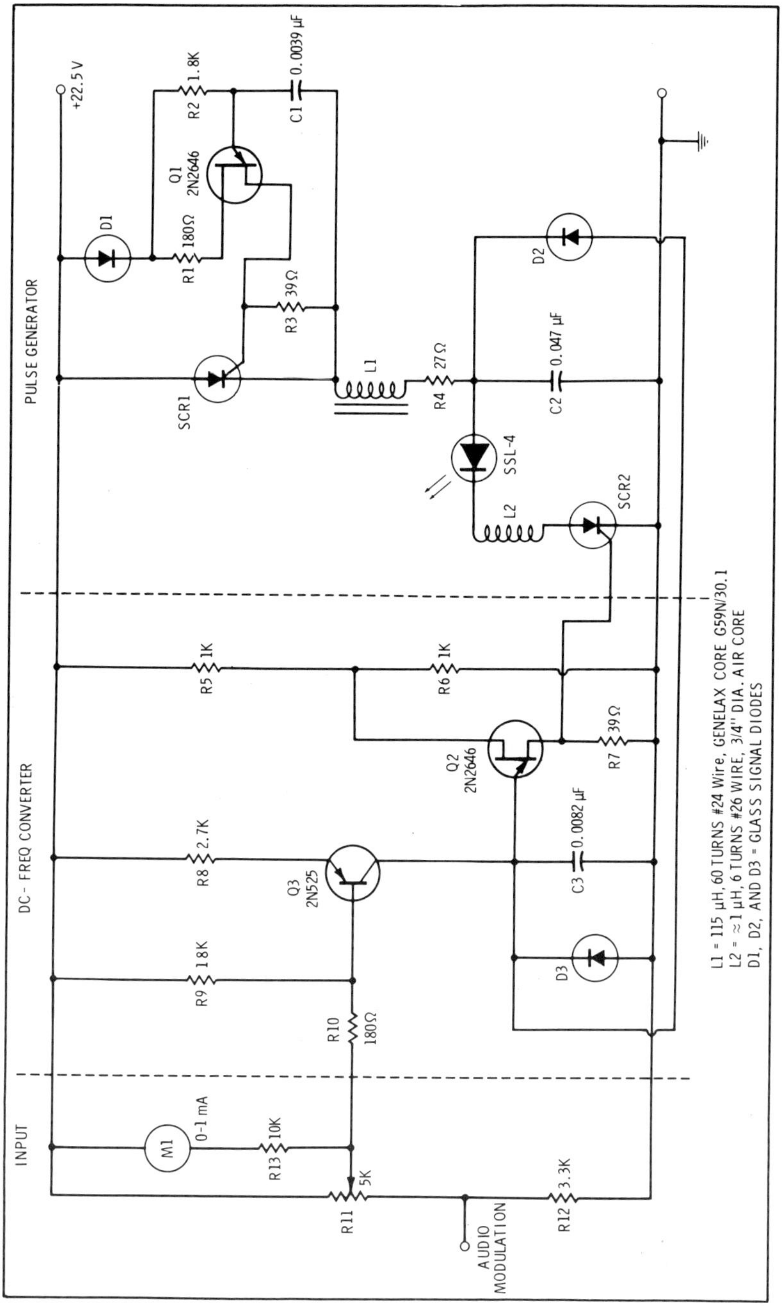

Fig. 5-19 (A) General Electric PM transmitter.

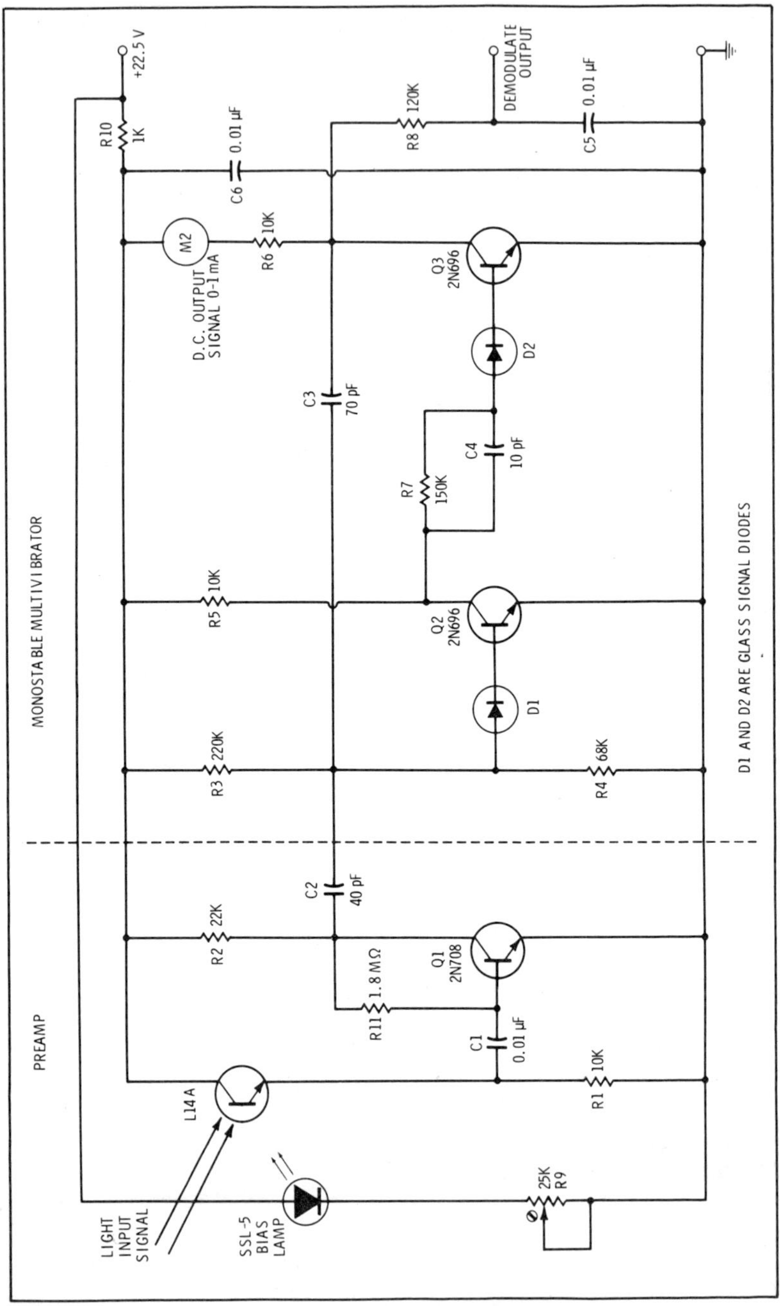

Fig. 5-19 (B) General Electric PM receiver.

crystal earphone. The 0- to 1-mA meter, which monitors the output signal, can be omitted from the circuit by connecting the 10K series resistor directly to the +22.5 volts.

The General Electric circuits can be used as a pair to form an optical transceiver, or either can be used with other prm equipment for a variety of applications. See *Solid State Lamps* for details on this and several other circuits.

SIMPLE PULSE-MODULATED TRANSMITTER

A very simple pfm transmitter is shown in Fig. 5-20. In operation, an incoming audio signal is amplified by Q1. The emitter of Q1 is connected to C1, the charging capacitor in a unijunction transistor relax-

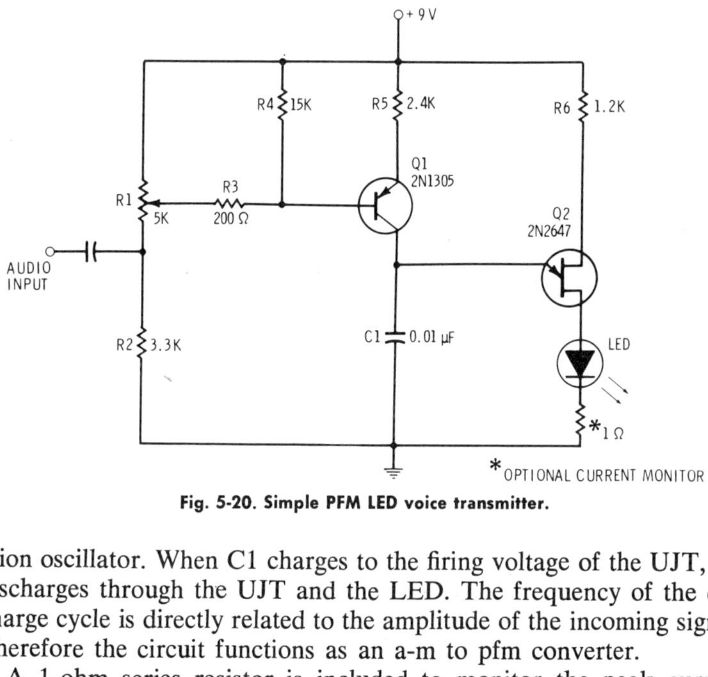

Fig. 5-20. Simple PFM LED voice transmitter.

ation oscillator. When C1 charges to the firing voltage of the UJT, C1 discharges through the UJT and the LED. The frequency of the discharge cycle is directly related to the amplitude of the incoming signal. Therefore the circuit functions as an a-m to pfm converter.

A 1-ohm series resistor is included to monitor the peak current through the LED. Like the previous pfm transmitter circuit, adjustments of R1 will be necessary to obtain optimum operation. If the charging time for C1 is too long, the carrier frequency will be too low for adequate voice reproduction. Too high a setting will cause overmodulation. Intermediate settings of R1 should cause a good quality tone which can serve as a marker signal when aligning a transmitter and receiver. The tone can also be used for sending coded transmissions. For best results, add a second potentiometer to the circuit in parallel to the first. The value of the new pot should be identical to the existing R1. A simple switching circuit will permit the two pots to be alternately connected into the circuit. For voice transmissions, leave R1 in the circuit. For an alignment tone, switch in the second pot. By using a pot for the additional resistor, the alignment tone can be varied.

The receiver circuit shown in Fig. 5-19(B) can be used to detect, amplify, and demodulate the signal from the simple pfm transmitter. Alternatively, any of the receiver circuits shown in this chapter can be easily modified for pfm operation by simply adding a demodulator circuit consisting of a monostable multivibrator. Fig. 5-21 shows a typical multivibrator demodulator circuit and both input and output connections. Integrated multivibrators are available for a few dollars, and they are much smaller than circuits using discrete components. While an IC can be used, the discrete circuit shown in Fig. 5-21 is

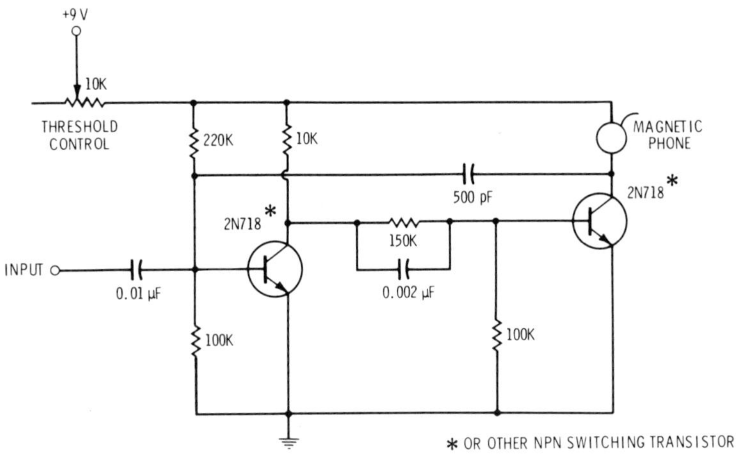

Fig. 5-21. Sensitive monostable multivibrator circuit.

more sensitive than IC versions. Appropriate adjustment of the potentiometer between the positive supply voltage and the multivibrator circuit gives a threshold sensitivity of a few tenths of a volt. Multivibrators of the IC type can require several volts for triggering.

The resistor-capacitor network at the output of the multivibrator averages the pulses from the multivibrator into an audio-frequency signal. Varying the values of either or both components will alter the tone quality of the output signal. Sometimes the network is not needed for acceptable demodulation.

The audio signal appearing at the output of the demodulator can directly drive magnetic phones. Volume may be reduced when the transmitter is set for optimum modulation, however, so a one- or two-stage amplifier may be necessary. The amplifier can be directly connected to the output of the demodulator.

LED-LED COMMUNICATORS

As was noted in Chapter 1, LEDs can be used to *detect* as well as emit radiation. A novel optical communication system, employing an LED as both transmitter and detector, has been designed by the author. The system uses a single amplifier and a switching circuit. Normally, the LED is connected in the detector mode and the circuit functions as a receiver. When a push-to-talk switch is pressed, the LED is connected to the output of the amplifier and the circuit becomes a transmitter. Fig. 5-22 is a block diagram for the LED-LED transceiver.

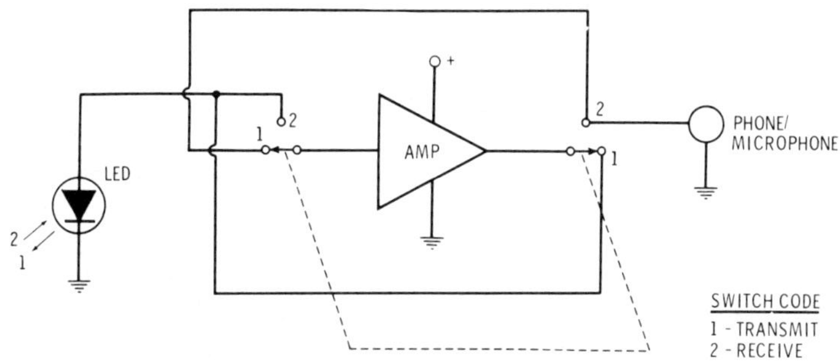

Fig. 5-22. LED-LED light-beam communicator.

The advantages of using a single optical source/detector in an optical communicator readily become apparent when an actual circuit is assembled. Since only a single lens, reflector, or optical fiber is required, there is a considerable space savings. In the case of a lens or reflector, the entire front surface of the communicator housing can be used while systems using a separate emitter and detector require two lenses or reflectors. This space savings doubles the potential collection area of the optics and increases communicating range.

Another advantage occurs from the use of identical optical systems at both ends of a link employing LED-LED transceivers. This means more of the collected radiation will be focused upon the active area of the LED being operated in the receive mode.

OPTICAL MULTIPLEXING

Multiplexing techniques can be used to transmit more than one channel of information over the same communications link. Several sophisticated techniques are available for both radio and light-beam multiplexing, but one of the simplest is a frequency-selective optical multiplexing scheme that permits several channels of information to be sent through a single fiber-optic strand. The technique employs LEDs with different wavelengths. In operation, each LED is coupled to the fiber simultaneously by means of secondary fibers or optical methods. At the receiving end, similar methods are used to split the beam into several paths, and an optical filter at each output passes only the desired wavelength. The result is that several different LEDs with various wavelengths can simultaneously transmit data over a single optical fiber. In its simplest form, a multiple-color LED could be employed as a light source.

Other optical multiplexing techniques are also available. For example, pulse modulation can be used to encode two or more signals on a single light beam. These schemes are valuable in some applications, but none are as simple as the multiple-color technique.

CHOOSING LEDS FOR COMMUNICATORS

Several companies market efficient GaAs and GaAs:Si LEDs suitable for use in optical communicators. For high-frequency systems requiring a fast response GaAs LED, the G.E. SSL-54 is a good choice. This diode typically emits one milliwatt at 100 mA forward current.

Voice communications require no more than a 10-kHz bandwidth for acceptable results, thus making high-power GaAs:Si diodes the best choice. Several good diodes are available, and one of the best is the SSL-55C. This LED typically emits 6 mW at 100 mA and higher values seem to be the rule. At one-ampere peak current, this diode emits about 38 mW.

Another good infrared LED is the TIL31. This diode typically emits 6 mW at 100 mA. For high-power communicators, the TIL27 is a good choice. This efficient emitter produces at least 15 mW at 300 mA continuous forward current.

For very high-power communicators, choose a hemispherical diode. The TIXL14 emits at least 60 mW at 300 mA and the TIXL16 emits at least 200 mW at 2 amperes. Unlike most planar emitters, a parabolic reflector can be used to capture most of the emission from these diodes. Unfortunately, the diodes are difficult to make and are therefore quite costly.

An LED designed primarily for use in optical card readers, the SG1004, may also find application in optical communicators. This little diode cannot be biased at the high-current levels of the other devices described here, but it is by far the most efficient. Conservatively rated at a typical output of 2.4 mW, this diode generally emits more than twice this value. Therefore, it may emit the same power as a higher current LED at only half the power input.

The high efficiency of the SG1004 is due to an efficient parabolic reflector which collects virtually all the radiation emitted by the LED chip and projects it into an external beam. Divergence of the beam is broad (half the power is contained within a 30° cone), but an external reflector or lens can be used to collimate some of the radiation into a narrow beam.

Table 5-1. Infrared LED Power Measurements

Diode	Manufacturer	Specified Output (mW)	Measured Output (mW)
SSL55CF	GE	5.4—7.5 @ 100 mA	6.52
SSL55CF	GE	5.4—7.5 @ 100 mA	6.74
SSL55CF	GE	5.4—7.5 @ 100 mA	6.74
TIL27	TI	6.7 typ. @ 100 mA	7.40
TIL31	TI	3.3—6.0 @ 100 mA	4.35
TIL32	TI	1.2 typ. @ 20 mA	1.74
SG1004	RCA	3.0 typ. @ 50 mA	4.57
SG1004	RCA	3.0 typ. @ 50 mA	5.87
SE5451	Spectronics	3.5 typ. @ 100 mA	3.91

NOTE: Power output measured with calibrated 2-cm × 2-cm silicon solar cell with sensitivity of 0.46 mA/mW.

Table 5-1 shows the power output from several commercial LEDs supplied to the author by the respective manufacturers. Measurements were made with a calibrated silicon solar cell. Power values listed in the table are subject to change since manufacturers sometimes improve their devices.

MEASURING LED OUTPUT

Effective calculations of the range to be expected of an LED communicator require a knowledge of LED power output. Ratings supplied by manufacturers frequently vary considerably from actual power outputs. Also, the use of external optics significantly affects the total radiated power from an LED communicator transmitter.

Many commercial systems for measuring both visible and infrared LED output are available, and one of the most economical is the 60-230 Photometer, shown in Fig. 5-23, manufactured by Metrologic Instruments, Inc. This photometer is equipped with a removable detector assembly and is primarily designed for measuring the power output of helium-neon lasers. Removing the cover over the silicon detector, however, permits the device to be used for measuring LED power output.

The Metrologic Photometer contains a dc-coupled operational amplifier circuit, and frequency response ranges from dc to 20 kHz.

Power readings in microwatts or milliwatts are read from the meter on the unit after a simple calculation which considers the spectral wavelength of the source being measured and the calibration of the silicon detector.

Tektronix, Inc., a major manufacturer of oscilloscopes, has developed a high-quality, portable photometer/radiometer which is ideal for measuring LED power output. Designated the J16, the basic radiometer consists of an amplifier, digital output circuitry, and a range circuit. Five different light-sensitive probes are available, each of which is compatible with a socket built into the J16. Some of the probes contain built-in subtractive filters to give the 1-cm^2 silicon detector a flat spectral response curve (within 5%) and others have a filter which gives the photopic luminosity curve (within 2%).

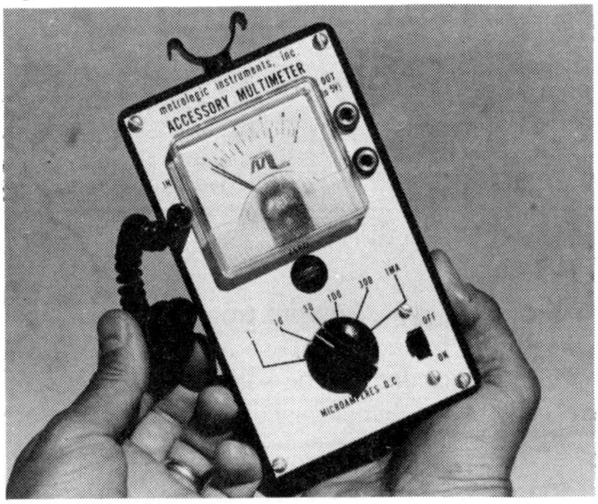

Fig. 5-23. Economical calibrated photometer.

Fig. 5-24 shows an engineer using the J16 to measure the output of an infrared LED. A precision Textronix power supply operated as a current source permitted accurate control over the forward current through the diode. Several of the LED output measurements given in this book were made with the apparatus shown in Fig. 5-23.

Optical power-measurement systems are also manufactured by Optronic Laboratories, Inc.; EG&G; United Detector Technology, and several other firms. Prices begin at about $60 for the Metrologic 60-230 and range to several thousand dollars.

No matter which measuring system is employed, the measurement accuracy is only as good as the instrument calibration. A poorly calibrated instrument can cause measurement errors of 100% or more. Fortunately, even a poorly calibrated instrument can be recalibrated for better performance. The typical procedure is to employ one of the standard lamp sources recommended by the National Bureau of Standards. Optronic Laboratories and EG&G sell calibrated standard lamps and regulated power supplies. Measurement instructions are also provided.

If a commercial power-measurement system is not available, an ordinary silicon solar cell and a 0- to 10-mA current meter can be used to measure the power output of infrared LEDs emitting more than a milliwatt continuously. For lower powers and for visible emitters, a 0- to 1-mA meter is required. Since the solar cell has a very linear response over a wide range of optical power densities, the cell

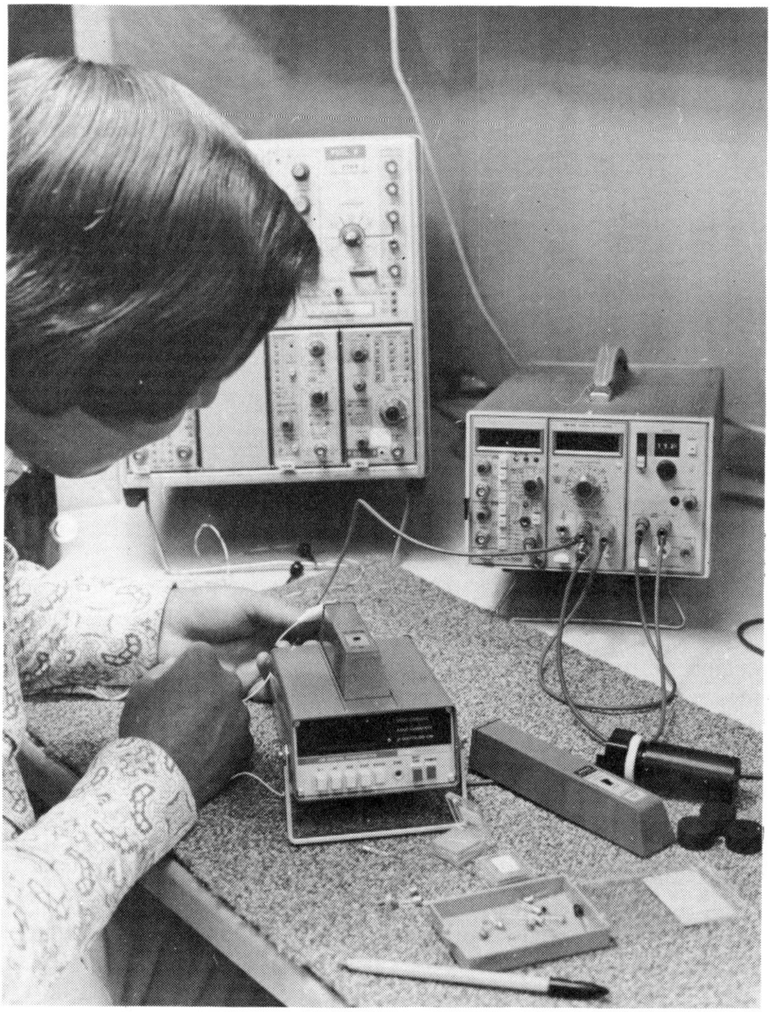

Fig. 5-24. Using 16 radiometer and required power supply to measure power output from LED.

and meter circuit need only be calibrated at one or two points for a single wavelength. Calibration, however, poses a major problem if a standard reference lamp or calibrated detector is not available. Typical silicon solar cells have a sensitivity of from about 0.4 to 0.5 mA/mW. That is, the cell generates a current of from 0.4 to 0.5 mA for every milliwatt of incoming radiation. These values are valid only at the peak spectral sensitivity of the cell being used, usually near 900 nm.

To illustrate the error potential of making infrared LED power measurements with an uncalibrated silicon solar cell, consider an LED which produces 1-mA current in a cell connected to a 0- to 10-mA meter. The 0.4-mA/mW sensitivity gives a power output of 2.5 mW, while the 0.5 mA/mW gives an output of 2.0 mW. The two readings differ by 20%, but it is likely that the average of the two measurements will fall within the power rating specified by the manufacturer. In fact, due to the uncertainty in the published ratings for LEDs, the measurement with an uncalibrated solar cell using an estimated sensitivity value will suffice for many purposes.

For very low power measurements, the current from the solar cell should be amplified. An operational amplifier is ideal for this role, and the op amp circuit used in the Tektronix J16 Radiometer is shown in Fig. 5-25. The amplifier can measure a wide range of optical power levels by means of a network of feedback resistors and a range switch.

When using an op amp circuit to amplify a solar cell current, be sure to isolate ambient light sources to prevent measurement errors. When very low powers are being measured, ambient light power can easily exceed the power of the source being measured.

MEASUREMENT PROCEDURES

Optical power measurement is not an exact science. Electronics engineers, technicians, and experimenters accustomed to making measurements of voltage, current, and resistance with better than 1.0% accuracy are usually shocked to find that optical power measurements with an accuracy of ±5% are rare and that ± 10% measurements are common. Even the National Bureau of Standards reports the calibrations it makes on standard lamps have a 4.1% uncertainty. Since all calibrated optical measuring equipment is traceable to the National Bureau of Standards, ultimate accuracy is hampered before a measurement is even made.

Major errors often occur during the measurement procedure itself. For example, *total* optical power cannot be accurately measured unless *all* the flux from the LED strikes the detector. The LED must either be placed directly adjacent to the detector or a reflector must be used to collect all stray radiation and reflect it toward the detector.

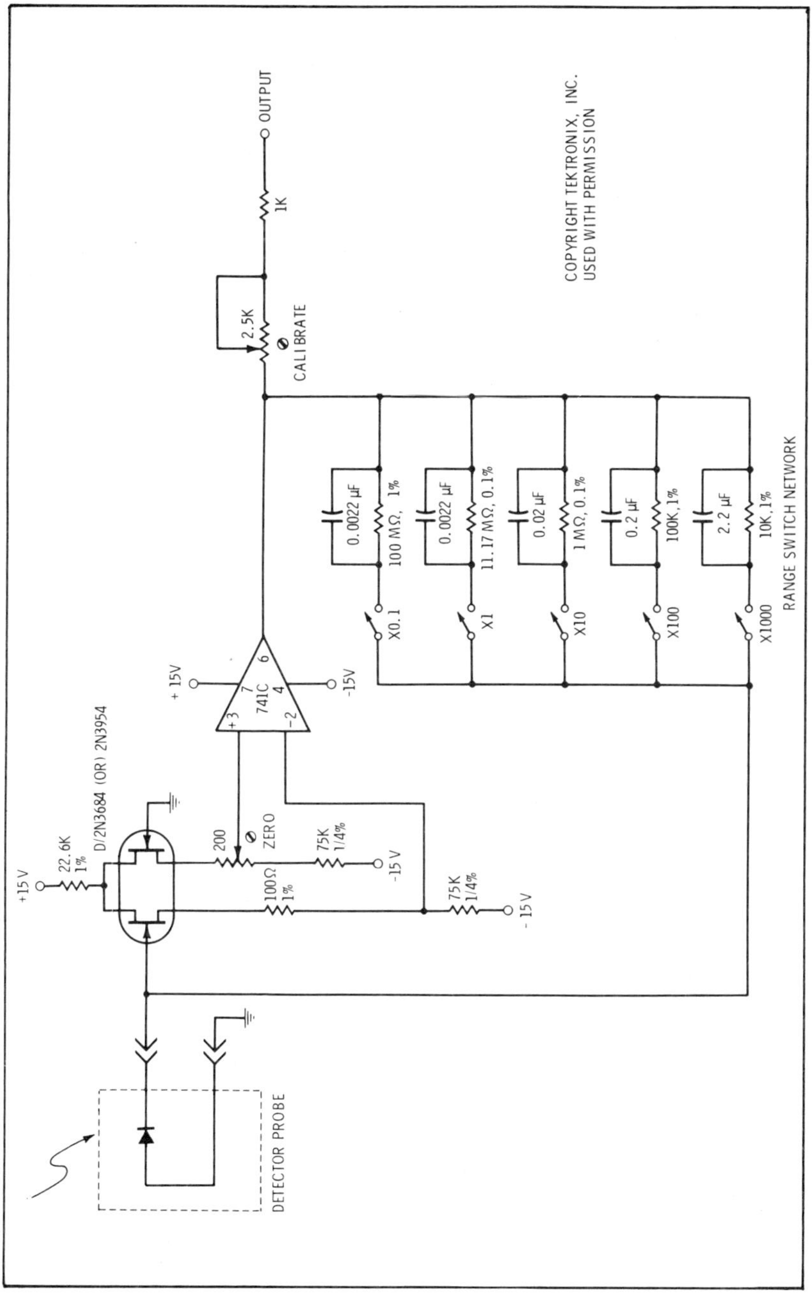

Fig. 5-25. Input section of J16 photometer.

In optical communication systems, total power emerging from an LED will not be collected by the lens, and in a typical system, only about one-fourth of the available light will be collimated into a narrow beam. If a small diameter lens is used, the power within the central beam can be measured by cutting a circle in a piece of black paper and placing the paper over a 2-cm by 2-cm solar cell. The transmitter is then moved near the cell so that only the central beam falls within the exposed area of the cell. If the transmitter is pulse modulated, an oscilloscope connected to a 100-ohm load resistor across the cell will be required to measure the LED power. The cell current will equal the voltage on the scope divided by the load resistance across the solar cell.

If a large transmitter lens is used, a power measurement near the center of the beam should be made. This reading can then be used to calculate the power density of the radiation within the beam. For example, 1.5 mW from 1 sq cm of exposed solar cell surface gives a power density of 1.5 mW/cm^2.

Summing up, use caution when measuring LED power. *Total* power output values provided by manufacturers can be of little value if the LED projects an off-axis halo.

OPTICS FOR COMMUNICATORS

External optics for LEDs are discussed in detail in Chapter 1. The information given there can be used to calculate the transmitted divergence of an LED transmitter.

In any optical system for LED communications, collection efficiency is of paramount importance. But remember that while a lens collects more of the radiation from an LED as it is moved closer to the diode, beam divergence increases.

For some short-range applications, no external lens or reflector is needed, and the integral optical system on the LED will suffice. Since the integral lens or reflector has a very short focal length, divergence of the emitted beam will be greater than if an external optical system is employed.

Parabolic reflectors usually give a higher collection efficiency for LEDs. Planar LEDs are best mounted *facing* the reflector surface. This causes some of the radiation to be blocked, but the overall efficiency of the reflector is still usually higher than a lens. In a test conducted by the author, an inexpensive flashlight reflector projected some 68% of the radiation from an SSL-55CF LED. The remaining 32% emerged from the opening normally occupied by the lamp installed in the back of the reflector. A good lens system collects only about half of the radiation from the SSL-55CF and projects only about 20% of the total radiation into a central beam.

Parabolic reflectors are available from the Edmund Scientific Company at a reasonable cost. Better-grade flashlight reflectors can also be used, and even a sealed-beam incandescent lamp can be adapted for use with an LED. The glass protective cover over the parabolic reflector should be carefully removed with a glass saw. Use great care to avoid physical injury during this operation. When the glass cover is removed, the LED can be soldered directly to the filament supports for the lamp. Make sure the LED chip is at, or very near, the focal point before soldering the diode in place. The filament is placed at the focal point, but infrared radiation will alter the focal length slightly.

CHAPTER 6

Intrusion Alarms and Ranging Systems

The discussion of source/sensor pairs in Chapter 3 described several interesting and useful applications for reflection and transmission sensors. By means of simple electronic circuitry and external optics, the operating range of reflection and transmission sensors can be increased substantially. Sophisticated intrusion alarms and detection systems are typical applications for these advanced source/sensor pair systems.

INTRUSION ALARMS

An LED intrusion alarm is essentially an advanced version of the source/sensor-pair reflection or transmission sensor described in Chapter 3. In the reflection mode, the beam from an LED is projected across a protected space such as a hallway or door opening. If a target with sufficient optical cross section penetrates the protected space and is illuminated by the LED beam, some of the radiation is reflected into a sensitive detector, where an alarm is triggered. The reflection mode is ideal for short-range intrusion alarm applications, but it is only practical for the protection of confined spaces which force the intruder to enter a relatively narrow opening.

A transmission-mode LED intrusion alarm is equivalent to a break-beam detection system. In this mode, a beam of infrared from an LED is directed toward a distant receiver. If the transmitted beam has a narrow divergence, a transmission-mode intrusion alarm can have considerably more range than a reflection-mode sensor system.

A transmission-mode system can be operated in a modified reflection mode. A retroflective target such as a highway reflector is mounted at one end of a space to be protected, and the LED transmitter is aimed at the reflector. The beam is then reflected back to a receiver, placed near the transmitter, which detects it. If the beam is penetrated, an alarm is triggered. A retroflector is used in this configuration to ease alignment. All that is required is that some of the beam strike the reflector and that the transmitter be adjacent to the receiver. For greater range, or for protecting a wider area, mirrors can be used. To protect a building, for example, mirrors at three corners reflect the beam from corner to corner. Alignment of a mirror system can be difficult, and this topic will be discussed later.

Both transmission and reflection intrusion alarms can use continuous- or pulse-operated LEDs. Pulse operation is almost always superior, and the reasons are similar to the advantages of pulsed frequency-modulated optical communicators over amplitude-modulated systems. A continuous-mode system requires high average power and is subject to interference from artificial and natural light sources. A pulsed system is relatively immune to the effects of other light sources and can generate high peak power from the LED illuminator. Continuous mode systems can be easily defeated with nothing more than a flashlight. The intruder simply illuminates the receiver with the light until he has penetrated the LED beam. Pulsed systems are far more difficult to defeat.

SIMPLE NONPULSED INTRUSION ALARM

A basic nonpulsed LED intrusion alarm is shown in Fig. 6-1. In operation, infrared radiation from a GaAs:Si LED falls upon the sensitive surface of a silicon solar cell. The op amp amplifies the output signal from the cell and closes a relay. When the LED radiation is blocked, the relay opens and an alarm is sounded. Both latching and nonlatching operation are possible, but the latching mode requires a reset switch.

While this basic nonpulsed intrusion alarm is very simple and low in cost, it does not have the range capability of pulsed circuits since

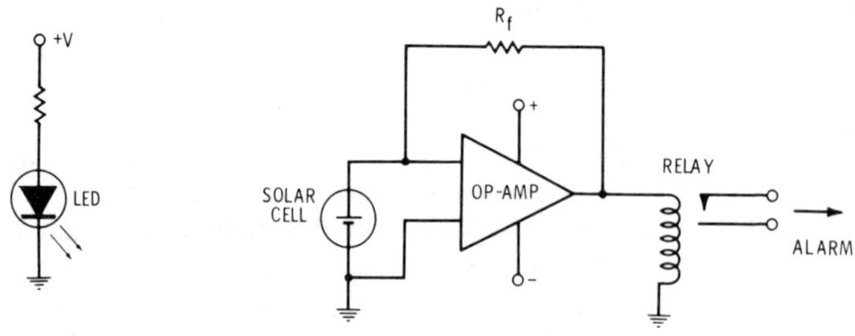

Fig. 6-1. Simple dc intrusion alarm.

the LED cannot be operated at high current levels. Also, the receiver is subject to interference from ambient light sources.

PULSED INTRUSION ALARM

The most important advantages of the LED nonpulsed intrusion alarm over conventional systems employing an incandescent lamp, include a totally invisible beam, low power consumption, miniaturization, and small optics. Much better performance can be achieved by pulsing the LED instead of operating it continuously. Pulsing permits more current to be injected into the LED, with a resultant increase in power output. Pulsing also permits the intrusion alarm receiver to be ac coupled and thus reduces the effects of ambient light. Finally, a pulsed system can be encoded to prevent an unauthorized bypass of the system.

General Electric has devised a simple but effective pulsed intrusion alarm which makes use of a simple two-transistor transmitter and a threshold receiver with self-contained alarm. The entire system will operate from flashlight batteries, so the alarm cannot be deactivated by turning off power lines. The circuitry is reproduced here with the permission of General Electric.

Fig. 6-2 shows the transmitter circuit of the intrusion alarm. The unit drives an LED with 50-microsecond pulses about one ampere in amplitude. The repetition rate is about 10 pulses per second, so the average current consumption is only 0.7 mA.

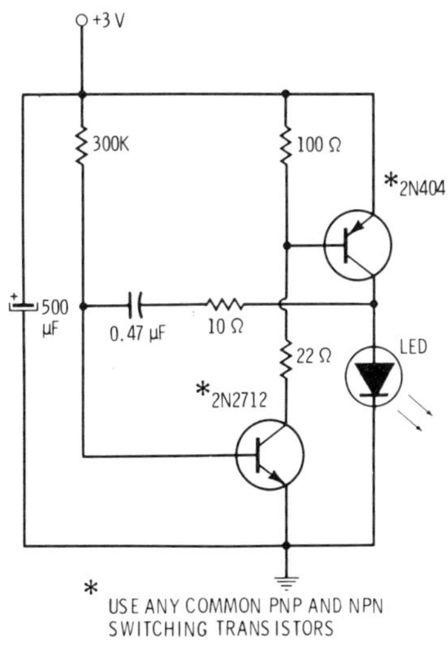

Fig. 6-2. Pulsed intrusion alarm transmitter.

Courtesy General Electric

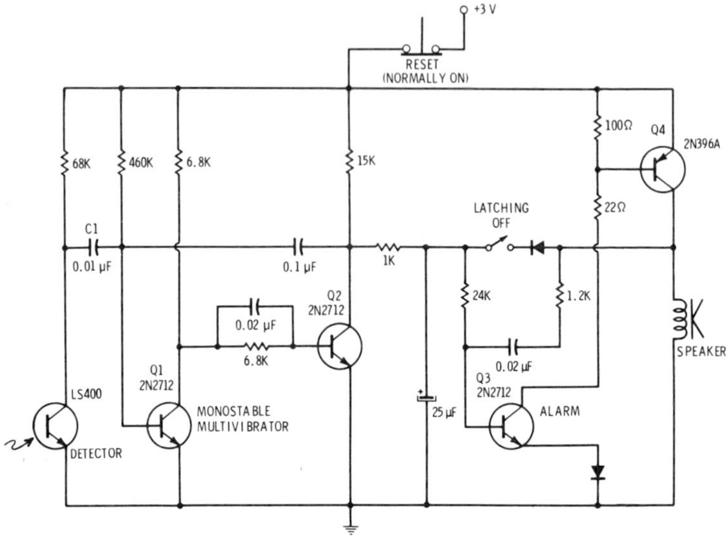

Courtesy General Electric

Fig. 6-3. Pulsed intrusion alarm receiver.

The intrusion alarm receiver is shown in Fig. 6-3. In operation, pulsed signals from the LED strike the sensitive surface of the photodiode and trigger the monostable multivibrator composed of Q1 and Q2. Ambient light is blocked by coupling the photodiode to the one-shot through a 0.01-μF capacitor. So long as pulses are received by the oscillator formed by Q3 and Q4, the alarm is not triggered. When the oscillator does not receive input pulses due to blockage of the LED pulses, the oscillator formed by Q3 and Q4 begins operation and signals a warning signal via the speaker.

In the latching mode, the alarm will deliver a warning signal until power is momentarily disconnected and the transmitted beam again allowed to strike the photodiode. When the latch is not connected, the alarm will sound only when the beam is blocked.

Though this alarm system is designed to operate in the presence of ambient light, excessive light at the sensitive surface of the photodiode will cause saturation and system failure. To avoid this possibility, place a light baffle or shield over the photodiode. Use of an infrared filter should also be considered.

LONG-RANGE INTRUSION ALARM

The circuits presented thus far have insufficient sensitivity for long-range applications. The receiver shown in Fig. 6-3 can be easily boosted in sensitivity by adding a high-gain amplifier between the detector and monostable multivibrator. One possibility is shown in Fig. 6-4. This circuit is inserted in place of C1 in Fig. 6-3 and has very high gain and is suitable for long-range detection. For even better results, substitute a pin photodiode, such as the EG&G SGD-040, for the LS400. The new diode should be *reverse biased;* that is, the anode should be connected to ground. Also, increase the photodiode load resistor to increase sensitivity. The best way to accomplish this is to substitute a one-megohm potentiometer and adjust for best results. Measure the resistance of the potentiometer and substitute a fixed resistor of the appropriate value.

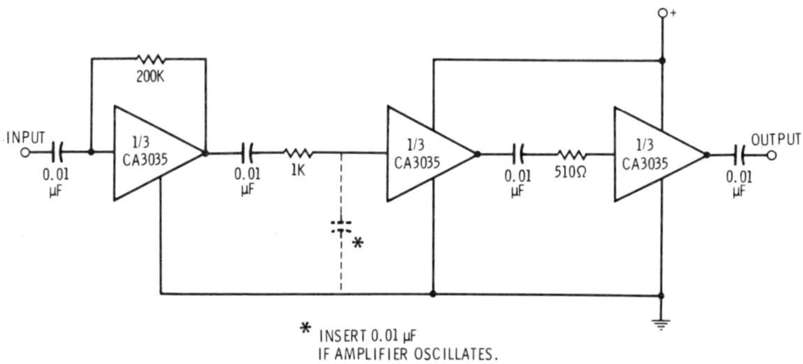

Fig. 6-4. High-gain amplifier for intrusion alarm receiver.

INTRUSION ALARM ALIGNMENT

Short-range intrusion alarm applications such as protection of a doorway require no complicated alignment procedure. The transmitter is simply aimed at the receiver, and both are mounted in a fixed position. Alternatively, the transmitter and receiver can be mounted in the same housing and a reflective target placed on a wall across the area to be protected. This technique makes concealment easier.

A system of mirrors that can be aimed is necessary to protect houses, offices, and other large structures or spaces. Many methods are possible, but it is important to align the mirrors carefully and mount them rigidly. If the mirrors move when wind strikes them, the alarm will be triggered.

Alignment of a mirror system can be difficult, and the best procedure is to employ an assistant. While one person adjusts the intrusion alarm transmitter so that its beam strikes the first mirror, the second person verifies the beam location with an infrared image converter or portable optical receiver. When the transmitter is aligned so that the first mirror is at the center of the beam, the process is repeated

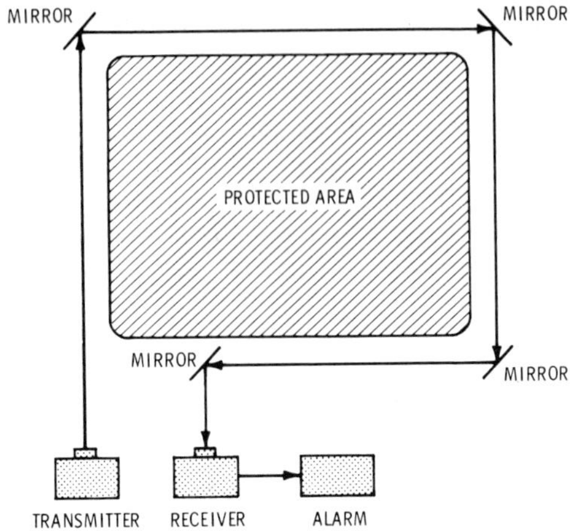

Fig. 6-5. Intrusion alarm mirror system.

for the first and second mirror until all mirrors are aligned and the transmitter beam strikes the receiver. A typical mirror system in proper alignment to detect intruders entering a large protected area is shown in Fig. 6-5.

LED RANGING SYSTEMS

Diode lasers and GaAs:Si and GaAs LEDs have been used in a variety of ranging systems, the most sophisticated being an optical radar that measures the time an infrared beam takes to travel to an object at an unknown range. In operation, an LED or laser emits a high-power pulse of radiation. A beam splitter reflects a small amount of the infrared toward a pin photodiode and causes the diode to activate a clock circuit. If its amplitude is sufficiently high, the receiver detects the reflected pulse and stops the clock circuit. Dividing the total time by two gives the one-way distance to the target.

Operation of an optical radar system requires sophisticated electronic circuitry, since receiver frequency response and transmitter rise time must be very high. Since light travels about one foot per nanosecond, sloppy rise times can introduce significant system errors. For very short ranges, the total time of flight may be less than 50 nanoseconds and an alternative ranging technique, optical triangulation, becomes feasible.

OPTICAL RADAR RANGE EQUATIONS

The maximum detection range for an optical detection system can be easily predicted with the help of several optical radar range equations. For a target that reflects diffusely and intercepts all of a transmitted beam, the maximum range for an active optical radar is

$$R = \sqrt{\frac{P_o A_r \rho \tau}{P_{th} \pi}} \qquad \text{(Eq. 6-1)}$$

where,

R is the range in meters,
P_o is the total transmitted power in watts,
P_{th} is the receiver sensitivity,
A_r is the receiver aperture area in square meters,
ρ is the reflectance of the target,
τ is absorption of the lens.

This range equation has been extensively tested by the author and found to be quite accurate for most applications. It is relatively simple to write a computer program for the equation and have a series of range graphs prepared for various system configurations. When a circular receiver aperture is employed, the equation can be simplified by expanding A_r into πr^2, where r is the lens radius. The two πs then cancel and leave

$$R = \sqrt{\frac{P_o r^2 \rho \tau}{P_{th}}} \qquad \text{(Eq. 6-2)}$$

Since the receiver area is a square function, doubling receiver diameter will double the detection range.

The equation can be used for purposes other than its original intention by solving for parameters other than range. For example, the reflectance of an unknown target can be experimentally measured by inserting the maximum range at which the target can be detected and solving for ρ. Similarly, P_{th}, a difficult parameter to measure, can be found by inserting the various values for each parameter and solving for P_{th}.

A second range equation must be used for a diffuse target that presents a cross section *smaller* than that of the transmitted beam:

$$R = \sqrt[4]{\frac{P_o A_t A_r \rho \tau}{P_{th} \Omega_o \pi}} \qquad \text{(Eq. 6-3)}$$

where,

A_t is the target area in square meters,
Ω_o is the transmitter beam divergence in radians.

Equation 6-3 is not as reliable as Equation 6-2 because of the nonuniform power distribution in a transmitted light beam. The equation is acceptable for general applications, but for precise uses it should be refined with a term defining power distribution in the transmitter beam. In addition to these basic range equations for diffuse targets, there are several for specular reflectors.

Remember that the range equations are only as accurate as the parameters plugged into them. If the transmitted beam is surrounded by a diverging halo, for example, the near range calculations will be less than experimental results, since the power at the target has increased.

While most system parameters can be obtained with little difficulty, target-reflectance values can be troublesome. Target reflectance can be measured using a simple device described later in this chapter, but many targets can have both specular *and* diffuse reflection characteristics, with resulting errors in range predictions. A good example is a waxed, wood panel. While the wood may be a relatively diffuse reflector, the wax will be a specular reflector because of its smoothness. Actually, even the wood will have a specular reflectance component.

The best compromise for a diffuse-specular target is to use diffuse reflectance values only. Since the specular component will only be noticed at the normal or straight-on viewing angle, diffuse reflectance will give a more representative prediction of performance.

Since target reflectance characteristics are so important to the detection range capability of an LED range finder, the topic will be discussed in more detail in the following sections.

TARGET CHARACTERISTICS

Depending on the intended application, an LED range finder may be required to detect a wide variety of targets. Some applications are less critical than others, and a simple system designed to detect trucks approaching a loading dock may be designed to detect a retroreflective tape such as Scotchlite, an almost ideal target. Other applications are more critical. An electro-optical aid for the blind, for example, is required to detect virtually any target in a wide range of ambient light conditions. Due to the diverse nature of targets it is designed to detect, a mobility aid for the blind represents one of the most demanding applications for an LED range finder.

In any range-finder application, a knowledge of target characteristics is essential. Of course, a fundamental feature of a target is its surface area. Targets which intercept only a portion of the beam, transmitted by the range finder, return less radiation to the receiver than targets intercepting all the beam. Any perforations which may be present must be subtracted from the total cross-sectional area of a target. Also, target area must be related to frontal cross-sectional area and *not* surface area.

The second most important target characteristic is the nature of its reflectance. As noted in the section on optical radar range equations, a target may be either a specular or diffuse reflector, and some targets have the characteristics of both. A specular target is a smooth, shiny surface which reflects an oncoming beam at the angle of incidence. The specularity of a surface is determined by the wavelength of the oncoming beam and the texture of the surface. If the surface texture is irregular and coarse, the oncoming beam will be scattered outward in a broad beam. If the surface is smooth, however, the beam will be reflected at the angle of incidence. Most specular reflectors are not perfect, and the reflected beam has a wider divergence than the oncoming beam. A nearly perfect specular reflector, such as a very flat front surface mirror, preserves the divergence of the oncoming beam in the reflected beam.

Diffuse reflectors have a rough, irregular surface. The surface may not appear rough to the touch or the unaided eye, but very small imperfections are required when the only specification is that they be larger than the wavelength of the oncoming beam. Paper, for example, usually appears very smooth, but paper is generally a nearly perfect diffuse reflector. To see why, examine a small square of paper under a microscope and notice its complex structure.

As noted earlier, many targets have both specular and diffuse reflection characteristics. The target described earlier, a waxed, wood panel, is essentially a sandwich of two targets. The wax layer forms a good specular reflector, while the wood forms a good diffuse reflector. Most leaves have a waxlike covering, and they are also good examples of specular/diffuse reflectors.

Other targets may have specular/diffuse reflectance without the need for two or more layers. Paper, for example, is coarse and porous and offers many different scattering surfaces. But paper also contains many firm, solid areas with a very smooth surface, particularly high-quality glossy paper intended for printing. Therefore, some kinds of paper may have both diffuse and specular characteristics. This phenomenon can be verified by observing a clear space on the outside cover of this book. Notice how a bright light source is reflected from the shiny paper. The inside cover is not finished to the same degree as the outside and therefore has little specular reflectance.

Diffuse and specular reflectors are illustrated in Fig. 3-16. Targets with both diffuse and specular reflectance are commonly described as having *gloss,* and an instrument called a *glossometer* is used to measure the spatial distribution of the reflected radiation.

From Fig. 3-16, it is obvious that the detection range of an LED range finder is limited by the nature of target reflectance. A specular target, for example, can return a very high percentage of an oncoming beam to a range-finder receiver, but only if the plane of the target is normal to the beam path. Otherwise, the beam will be reflected away from the receiver. A diffuse target will always reflect some oncoming radiation back toward the receiver, but since the radiation is scattered into a very wide pattern, very little radiation actually strikes the receiver.

A *cooperative target* such as a retroreflector is the best solution when a range finder is to detect a specific object to which a retroreflector can be attached. High-quality glass cube-corners, or corner reflectors are the best choice, and scientific suppliers such as Edmund Scientific and Metrologic Instruments stock several different types. Clear-plastic highway reflectors are a far more economical choice, but they reflect the oncoming beam at a wider divergence than glass retroreflectors. For short-range applications Scotchlite, a thin, rugged tape, coated with tiny glass beads, can be used. Also, fine glass beads of the type mixed with paint for marking pavement and traffic signs can be applied to the target. For best results, paint the surface to be coated and, while the paint is still wet, spray or strew the beads across its surface.

Most LED detection systems must be capable of detecting a great variety of targets, most of which will be diffuse reflectors with perhaps a slight degree of gloss. Calculations of predicted detection range are best made by assuming diffuse reflectance. The actual reflectance of the target must then be determined, and this topic is discussed next.

TARGET REFLECTANCE

With the exception of target reflectance, most of the parameters to be inserted into optical radar-range equations are not difficult to determine. Some researchers have investigated the near-infrared reflectances of a variety of materials and have published their findings. Unfortunately, there have been comparatively few studies of this nature, and some of the studies contradict one another.

To permit accurate predictions for the expected range of an LED detection system for the blind, the author has designed a simple, low-cost reflectometer that can be used to measure the reflectance of diffuse surfaces at 940 nm. The device, which is shown schematically in Fig. 6-6, consists of a recessed detector head containing an SSL-55C GaAs:Si LED surrounded by a series-connected circular array of six 2- by 4-mm silicon solar cells. The detector head is made of an opaque material to prevent stray light from affecting the measurements. The solar-cell array is connected directly to a 0- to 50-microampere meter, and the LED is biased with a 6-volt lantern battery. The LED current is limited to 100 mA by a 0- to 100-ohm adjustable series resistor.

Before the reflectometer can be used, it must be calibrated. The prototype unit was calibrated with several materials with known reflectance values at 940 nm, traceable to at least two published references each. The calibration procedure consisted of three steps. First, an aluminum plate was coated with a layer of smoked magnesium oxide (MgO), in accordance with instructions provided by the National Bureau of Standards. The procedure can be duplicated by simply burning a magnesium ribbon held with metal tongs and allowing the smoke to form a thick, 1- to 2-mm layer on the aluminum plate. Smoked MgO has a reflectance of 98% at 950 nm. The reflectometer probe was gently pressed against the MgO sample, and the LED current was adjusted until the meter read 49 mA (or ½ of 98%). Next, the probe was placed against a flat sample of fresh, clean asphalt that had a known reflectance of 5%. The 50-microampere meter gave a reading of 2.5 μA, thus indicating a linear response for the solar cells. A third reading was made with a vinyl-coated nylon fabric with a known reflectance of 15%, and the meter showed a reflectance of 16%.

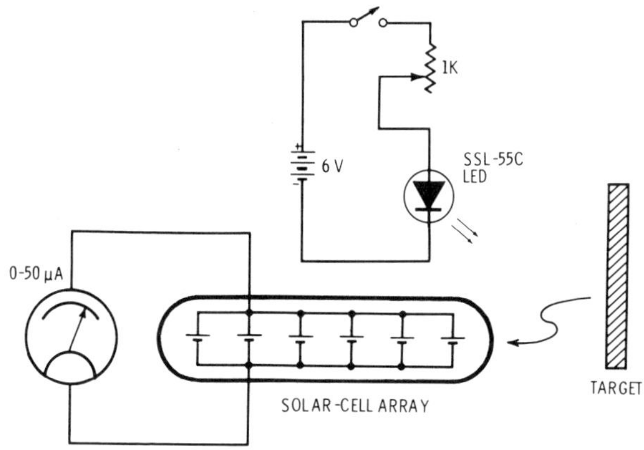

Fig. 6-6. Reflectometer circuit.

Next, the reflectometer was used to measure the reflectance of a thick, white paper card to be used as a secondary reflectance standard. Smoked MgO is very fragile, and the secondary standard provides a sturdy and quick way to calibrate the reflectometer before use. The card selected had a reflectance of 90%. By comparison, the paper in a book had a reflectance of 89% (as measured against a blank space on the title page without correcting for the reflectance contributed by underlying pages).

The reflectometer, shown in Fig. 6-7, has been used to measure the reflectance of hundreds of materials, and some of the results are presented in Table 6-1. The reflectance data can be used in calculating the range of an optical radar in some situations, but every target should be treated on its own merits. For example, a single leaf may have a reflectance of only 50%, while a *layer* of leaves may have an 85% reflectance. Similarly, depending on its composition, "gray" concrete may have a reflectance ranging from 35% to almost 60%.

Fig. 6-7. Author's wife using low-cost reflectometer to measure near-infrared reflectance of cottonwood bark.

Table 6-1. Typical Reflectances at 950 Nanometers

Material	Reflectance (%)
Tar Paper	5
Natural Rubber	5
New Asphalt	5
Corn Leaf*	45
Squash Leaf*	48
Silver Maple Leaf*	52
Aspen Leaf*	56
Prickley Pear Cactus	60
Brown Corrugated Board	75
Styrofoam	83
Cotton Fabric (Infinite Layer)	87
Powdered Chalk ($CaCO_3$)	94
Smoked Magnesium Oxide (MgO)†	98

*Upper surface of living leaf
†Calibration standard

OPTICAL TRIANGULATION

For short-range detection applications, an optical triangulation system can give reasonably accurate results with very simple equipment. Fig. 6-8 shows the geometry of a simple optical-triangulation scheme employing an LED or a diode laser. In operation, the beam from the diode light source is pointed toward a target at an unknown range. A receiver a fixed distance from the transmitter is then rotated until its field of view intercepts the illuminated spot on the target. The

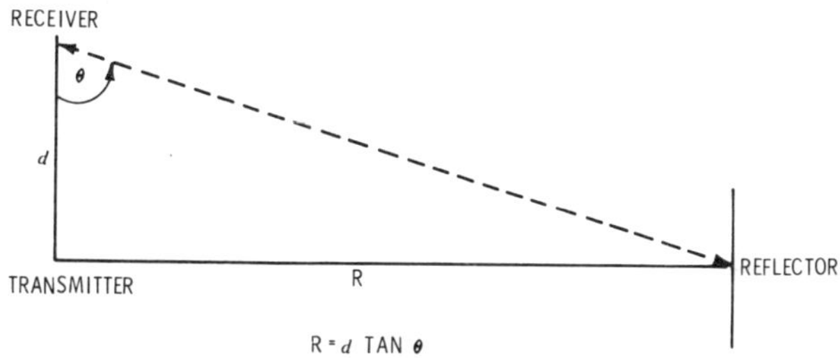

Fig. 6-8. Optical triangulation.

angle of intersection is measured and, together with the fixed distance from transmitter to receiver and simple trigonometry, is used to determine the range to the target. In a typical system, the range is given by

$$R = d \tan \theta \qquad \text{(Eq. 6-4)}$$

where,

d is the fixed distance from transmitter to receiver,
θ is the angle of intersection.

The resolution (accuracy) of a triangulation system is determined by the divergence of the transmitter and the field of view of the receiver, with greatest accuracy resulting when both are small. For more detailed information on optical triangulation, see W.J. Hannan, "Application of Injection Lasers to Communication and Radar System," *RCA Lasers,* pages 95-99.

RANGE-FINDER APPLICATIONS

Laser and LED range finders have found use in a variety of military, industrial, and even commercial applications. For maximum range, diode laser transmitters are preferred, but recently the availability of commercial avalanche photodiode modules has made possible moderately long-range systems using LEDs.

A potential volume market for low-cost LED triangulation detection systems is automobile and pleasure boat "radars." By means of a scanning receiver system and a fixed LED beam, range indications in increments of several meters could be signaled by an LED readout. This kind of system could be used to help space cars on the highway and avoid collisions in fog by small boats. In both cases, reflectors would enhance detection by increasing the strength of a returned signal. A similar system could be used at the rear of a truck to assist in backing up to loading docks. Alternatively, the "radar" could be mounted to the dock and the driver could be given the separation distance by a large indicator observed through a side mirror.

Besides measuring the range, LED range-finder systems can be used to count vehicles, pedestrians, manufactured goods, and other objects. In a vehicular-counting application, an LED transmitter-receiver pair is mounted on one side of the road and aligned so that the device will detect vehicles in the nearest lane of traffic. Each time a vehicle enters the detection region, a counter is activated. This same system can be used to detect trucks and buses but to ignore passenger cars by placing the detection region above the height of the average automobile.

An important commercial application for LED triangulation detection systems is electronic travel aids for the blind. As noted earlier, a travel aid for the blind must be capable of detecting a wide variety of targets. A minimum detection range of at least a meter is required, and the physical size and weight of the device must be kept as small as possible. Since a practical travel aid incorporates most of the components required for a simplified ranging system, a separate section will be devoted to describing a few working devices.

LED TRAVEL AIDS FOR THE BLIND

A typical low-cost electronic travel aid for the blind that detects objects with a triangulation technique is shown in Fig. 6-9. In operation, the unit projects a narrow beam of near infrared from a GaAs:Si LED in a movable lens-tube assembly. If an object within the detection range of the system is irradiated by the infrared beam, some of the radiation is reflected back toward a sensitive phototransistor in a second lens-tube assembly. The signal is amplified by a linear IC and passed on to an earphone in the user's ear. Since the transmitter is modulated by a 500-Hz tone, the user hears a tone when an object is detected.

The triangulation feature of the travel aid permits the user to obtain rough range information about a detected object. If a signal is heard when the transmitter lens tube is depressed (i.e., at an angle to the receiver tube), then the detected object is within a meter. If the transmitter lens tube is parallel with the receiver tube, then the detected object is greater than 3 meters away. Intermediate positions of the transmitter tube give other range indications.

The travel aid in Fig. 6-9 is inexpensive and operates well, but the requirement for hand-held use interferes with the user's freedom of movement. To help solve this problem, the author has installed an LED triangulation system for the blind in an eyeglass configuration. The result, shown in Fig. 6-10, is the smallest, optical-ranging system ever designed.

Operation of the eyeglass aid is similar to that of the larger hand-held unit. A transmitter system in a brass tube on one side of the glasses emits 90-mW pulses of near infrared (940 nm) from a GE SSL-55CF GaAs:Si LED at the rate of 120 Hz, and some of the infrared is collimated into a 0.09-radian (5-degree) beam by a simple

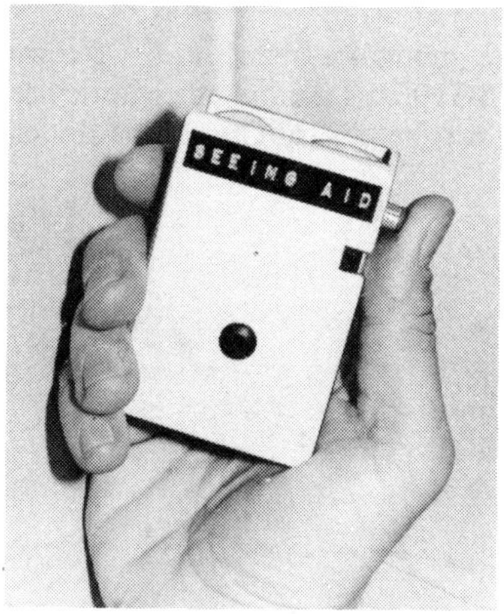

Fig. 6-9. LED travel aid for use by the blind.

double-convex f/1.2 lens. Fig. 6-11 is an infrared photograph showing the 940-nm radiation of the eyeglass aid. Infrared striking an object is scattered back toward the aid, and a small amount enters the 12-mm diameter receiver lens. The radiation is detected by a silicon pin photodiode and passed on to a high-gain integrated circuit amplifier. The signal is amplified 80 dB and presented to a monostable multivibration threshold circuit. If the signal amplitude exceeds the trigger level of the threshold circuit (about 0.3 volt), the single-shot multivibrator issues a constant-amplitude output pulse that is converted to an audible signal by a miniature hearing-aid receiver and passed to the user's ear via a thin plastic tube. The combined voltage gain of the amplifier and single-shot is 110 dB.

Though the LED emits 90 mW of near infrared, only half the radiation strikes the lens, and only about 17 mW of this is projected into a central beam. A diverging halo surrounding the primary beam contains 18.5 mW. An SSL-55CF LED is employed, since it incorporates

Fig. 6-10. LED travel aid for the blind mounted to eyeglass frames.

a miniature reflector for improved collection efficiency. The RCA SG1004 is a more efficient device, since there are no packaging losses, but the SSL-55CF can be driven with much higher-current pulses.

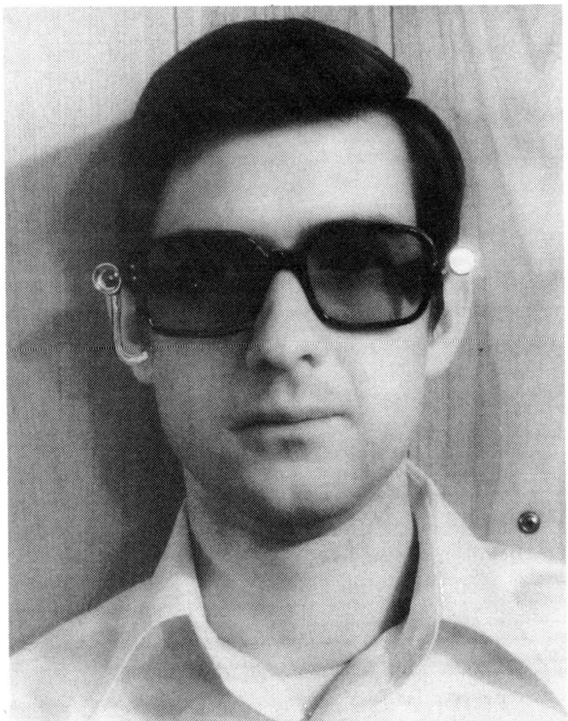

Fig. 6-11. Inrfared photograph showing eyeglass-mounted travel aid for blind. Note bright light from transmitter lens (right).

Fig. 6-12 gives the complete circuit diagram for the eyeglass travel aid. With the values shown, the aid has the electrical and optical specifications shown in Table 6-2. Operating power is supplied by a 6.8-volt mercury battery (Mallory TR175) in both the transmitter and receiver tubes. The transmitter battery has a life of 32 hours and the receiver battery has a life of 62 hours. Total operating cost comes to seven and one-half cents per hour.

The electronic circuits are assembled on miniature etched-circuit boards, one for the transmitter and two for the receiver (amplifier plus single-shot multivibrator). Fig. 6-13 shows the receiver boards. Acrylic bulkheads secure the circuit boards in place and provide contact points for the battery. A removable switch assembly at the rear of each electronics tube, each of which measures 13 by 90 mm, facilitates battery replacement. A disassembled view of the eyeglass aid is shown in Fig. 6-14. Design improvements will result in even further miniaturization and higher sensitivity.

Table 6-2. Electro-Optical Specifications for LED Detection System for Blind

Transmitter	
Peak Optical Power	950 nanometers
Wavelength of Emission	0.08 radian (4.6°)
Beam Divergence	20 microseconds
Pulse Width	120 Hertz
Pulse-Repetition Rate	6.8 volts (TR175)
Power Supply	5.5 milliamperes
Current Drain	17 milliwatts
Receiver	
Sensitivity (Optical)	20 nanowatts
Frequency Response	20 kilohertz
Rise Time	10 microseconds
Field of View	0.08 radian (4.6°)
Power Supply	6.8 volts (TR175)
Current Drain	3.0 milliamperes
Gain	95 dB (voltage)

Though the combined weight of the transmitter and receiver is less than 85 grams (3 ounces), the unit has a power out/power threshold ratio (P_o/P_{th}) of 9×10^5, and, with subsequent design improvements, this value will be raised to at least 2×10^6. Fig. 6-15 shows the range capability of the present aid as predicted by theory (solid line) and confirmed by experiment (circles). Note that the agreement between the predicted and experimental results is quite good. The dotted line on the graph shows the range predicted for an aid with a P_o/P_{th} of 2×10^6. The range predictions were made with the help of Equation 6-1 and apply to diffuse targets with a cross section which exceeds that of the oncoming beam.

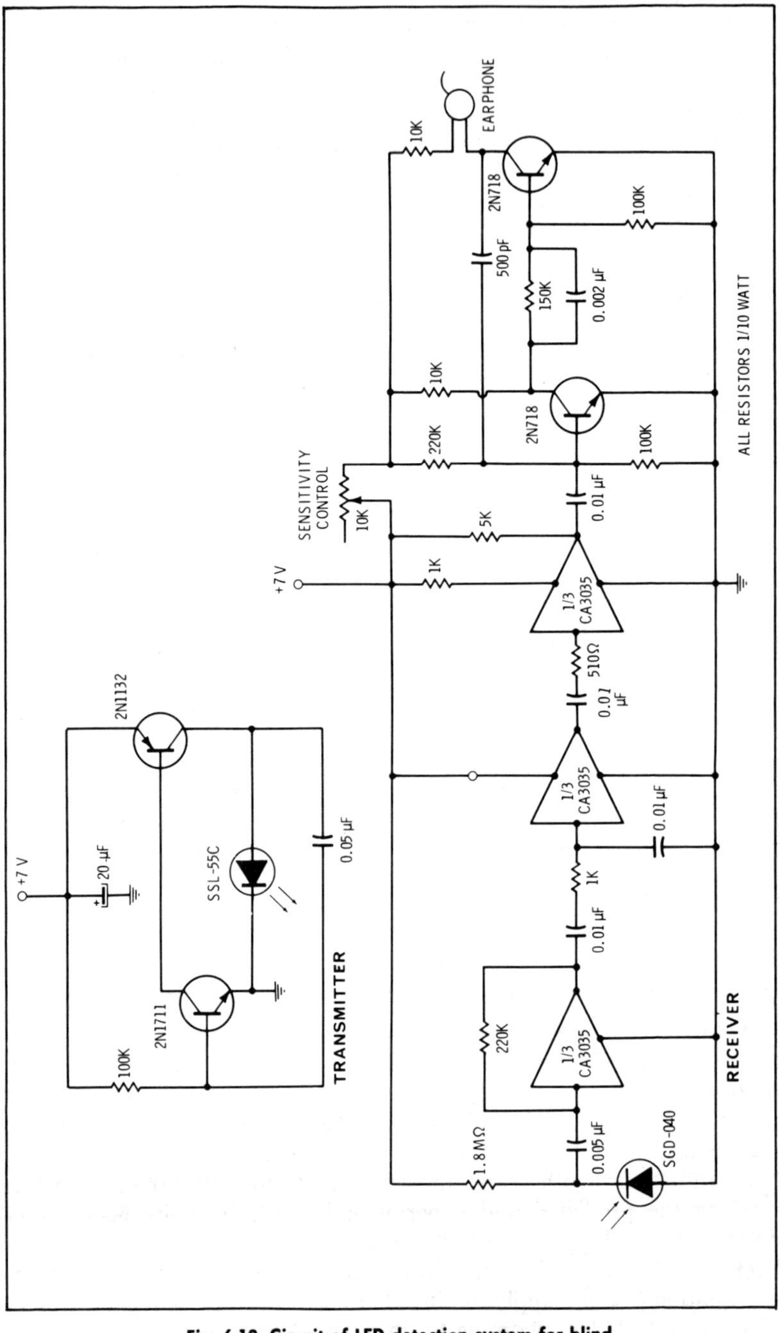

Fig. 6-12. Circuit of LED detection system for blind.

Fig. 6-13. Receiver circuit boards used in eyeglass travel aid (amplifier on left, threshold circuit on right).

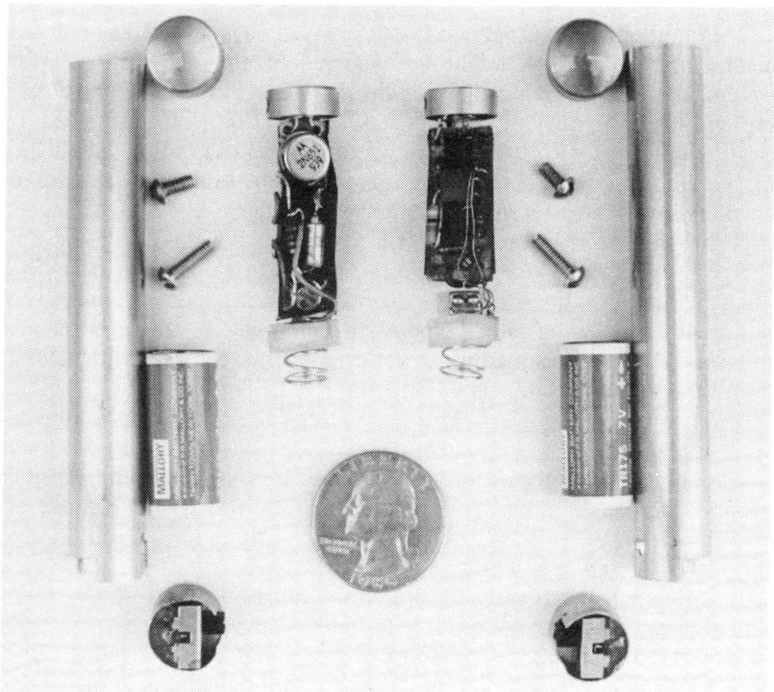

Fig. 6-14. Disassembled view of eyeglass travel aid.

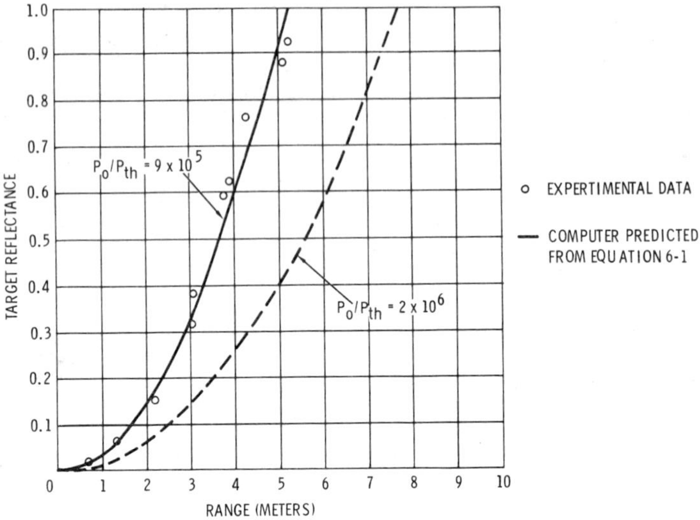

Fig. 6-15. Predicted vs. experimental range for LED detection system.

This miniature LED detection system has been tested in environmental extremes which might be expected to alter its performance. Detection range is reduced when the target is illuminated by direct sunlight, causing the onset of detector saturation, but this effect is easily eliminated with the addition of an optical filter. Fog caused no measurable change in detection range, and snow, rain, and hail caused an occasional false return in the form of solitary clicks. The clicks occur when a pulse of radiation strikes a single snowflake or raindrop.

Storage at −23 °C for eight hours produced only a slight detection range reduction, and storage at 46 °C for three hours produced no reduction in detection range. Operation in humidities ranging from 10 to 95% (relative) resulted in no noticeable detection range reduction, even though the 940-nm radiation from the LED falls partially within a water absorption region. The aid successfully withstood drop and vibration tests.

Though the circuitry of the aid is not affected by high humidity and the aid has been successfully operated in the rain, the reflectance of many substances is reduced by moisture. Therefore, the detection range can be reduced when targets are wet.

Since the aid, in its present configuration, can be used for cw communications via the transmitted tone, the device has other applications as well. In fact, a modification of the device has been developed that makes possible good-quality prm voice communications. These eyeglass infrared transceivers could be useful in a variety of situations requiring moderate range, line of sight, and secret communications.

The eyeglass electronic travel aid is now undergoing further development and testing and will eventually be made available commercially. The commercial aid will incorporate either two LED transmitters or two receivers to provide range information. In the case of two LED transmitters, the beams would intersect the field of view of the single receiver at ranges of one and three meters. The beams would be sufficiently divergent to overlap one another at about two meters. As a target is approached, the first signal would be a tone at the pulse-repetition rate of the first LED. The second signal would be a mixture of the modulation rates of both LEDs, and the third signal the modulation rate of the second LED. This method gives three distinct detection zones with only two transmitters and a single receiver. Similar results can be obtained with two receivers and a single transmitter.

Travel aids for the blind provide an excellent example of the miniaturization potential of sophisticated communications and ranging systems employing readily available, low-cost LEDs. In the future, even more diverse roles will be found for new and better light emitting diodes.

CHAPTER 7

The Injection Laser

Semiconductor light sources were considered as possible candidates for laser action soon after the invention of the first laser by Theodore Maiman in 1960. Efficient infrared emission in the gallium arsenide light emitting diode was reported in 1962, and several laboratory teams immediately began work to develop a semiconductor laser. In the fall of 1962, scientists at three laboratories almost simultaneously announced the invention of successful semiconductor lasers. General Electric was first; IBM and MIT were close behind. Thus far, no other laser has been developed which has higher efficiency.

The semiconductor injection laser is essentially a specially prepared LED; therefore, it is important to describe its operation and applications in a book on light emitting diodes. This is particularly true now that injection lasers are far more sophisticated and reliable.

The injection laser is a light emitting diode with a very flat junction and two end mirrors (Fig. 7-1). The mechanism responsible for the light generation in an injection laser is identical to that of the LED. Electrons from an external power supply are injected into the n-side of the junction. The electrons are excited to a higher-than-normal energy state, and, after crossing the junction, they fall into holes. This process creates energy in the form of photons and heat.

Below a critical point known as the *lasting threshold,* the injection laser emits light spontaneously and randomly like an LED. If sufficient current is applied to the device (a very large number of electrons are injected into the crystal), a situation occurs where there are more electrons in an excited than in an unexcited state. This condition is called a *population inversion* and is essential to laser action. A randomly emitted photon can then act to stimulate the emission of a photon from an excited electron. The new photon may do likewise, and the process continues in the form of a chain reaction. The process is called *stimulated emission of radiation.* The word "laser" is an acronym for Light Amplification by Stimulated Emission of Radiation.

The importance of the population inversion can be more readily understood if we once again consider light emission in an LED. Since the LED cannot support an inversion of electrons, emitted photons may be absorbed by unexcited electrons (perhaps stimulating them to a higher energy level). When a population inversion is present, photons are more likely to strike excited electrons and stimulate the emission of still more photons. The inversion then sets up a situation where optical gain exceeds optical loss.

Since the injection laser has been given two, parallel-facing end mirrors, some of the photons are reflected back into the active region

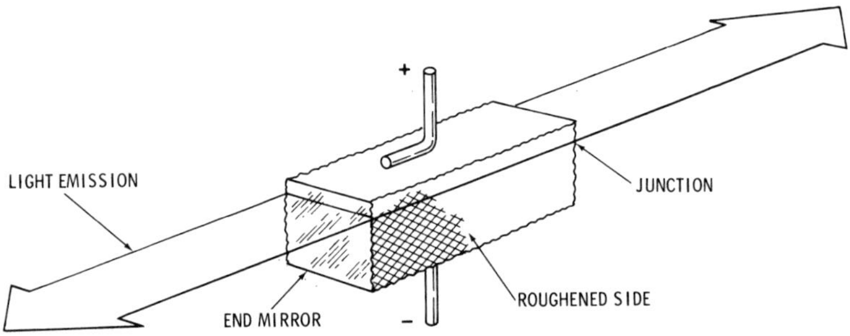

Fig. 7-1. Basic injector laser.

along the junction. There, they stimulate still more electrons into emitting photons and an oscillating wave of light is built up along the junction. The process is almost identical in principle to feedback in a resonant electrical circuit. Since the photons are "in step" with each other, the wave of light satisfies the requirement for phase coherence.

Some of the theoretical aspects of injection lasers are summarized in Fig. 7-2. Beginning at Fig. 7-2A, an injection laser, operated below the lasing threshold, acts like an LED and emits light randomly. As more current is applied (Fig. 7-2B), a point is suddenly reached where a population inversion exists, and more electrons are in an excited than in an unexcited state. Stimulated emission, the key to laser action, occurs in Fig. 7-2C as randomly emitted photons collide with excited electrons in the active region along the junction and cause the emission of additional photons. The new photons are emitted in phase with their progenitors. Finally, in Fig. 7-2D a standing wave of photons is set up between the two end mirrors, composing the optically resonant cavity necessary for laser action.

Unlike most other lasers, the end mirrors of the injection laser are an integral part of the device. The high index of refraction for GaAs discussed earlier may be a disadvantage for LEDs, but it is certainly important to the injection laser. GaAs is a crystal that may be cleaved along certain crystalline planes to produce extremely flat, parallel facets on opposite sides of a chip. These facets have a reflectance of approximately 35% at the GaAs wavelength of 903 nanometers and make excellent end mirrors.

The lasing threshold may be lowered by placing a 100% reflecting mirror over one of the two end facets. The mirror is often a thin film of gold, insulated from the junction by a layer of silicon dioxide to

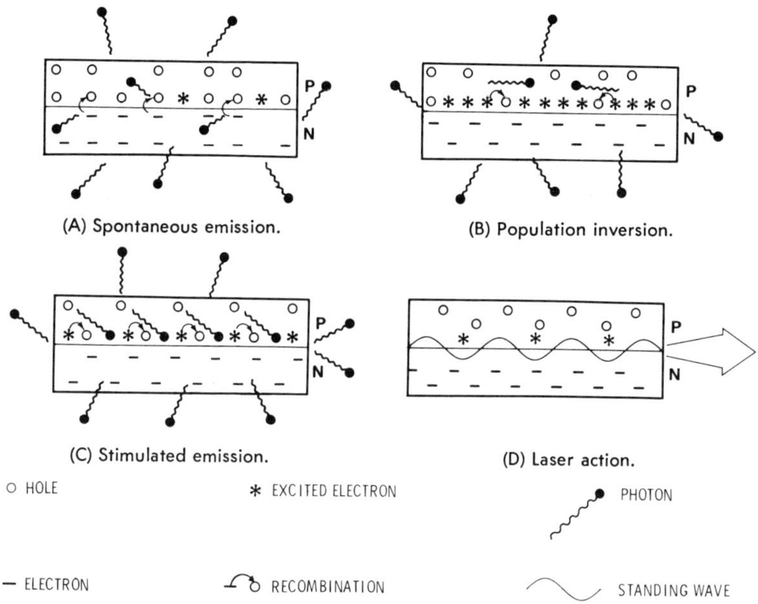

Fig. 7-2. Steps preceding laser action.

prevent an electrical short. Light that would have been expelled from the lasing cavity is then reflected back into the active region, where it contributes to gain by stimulating the emission of additional photons. A practical aspect of using the 100% end mirror is that it is much easier to make use of the total power output of a laser when it exits from only one end of the device. When light is emitted from both ends of a laser, half of it is either redirected in line with the primary beam by means of mirrors, or is lost.

LASER STRUCTURES

The first injection lasers were made by diffusing zinc into n-type GaAs to give a pn junction. Heat treatment displaced the junction slightly and formed an optical waveguide region. Formation of the waveguide significantly improved laser operation by reducing the required current density at the pn junction.

Unfortunately, these early diffused lasers required current densities of at least 60,000 A/cm^2 to reach the lasing threshold and well over 100,000 A/cm^2 for operation at reasonable power output levels. These high-current densities caused a degradation in output power; only by cooling the laser to cryogenic temperatures, with liquid nitrogen or liquid helium, could the threshold values be significantly reduced. The lower thresholds at cryogenic temperatures also permitted continuous operation. Due to heat caused by the very high-current densities, room-temperature operation was limited to pulse service.

In 1963 RCA developed a new method for forming pn laser junctions superior to the zinc diffusion process. The process, called liquid epitaxy, consists of wetting a polished wafer of n-type GaAs with molten p-type GaAs at a precise temperature. When the combination is cooled, the molten material has melted the first few microns of the substrate wafer. This crystalizes onto the wafer, forming a very flat, uniform junction. As with diffused lasers, heat treatment is employed to displace the junction and form a waveguide region.

Lasers made using the epitaxial process have significantly lower thresholds than diffused lasers. Typical threshold current densities range from 40,000 to 60,000 A/cm^2. Typical operating current densities range from about 120,000 to 180,000 A/cm^2.

RCA made another important contribution to injection laser technology with the development of the *heterostructure* laser. The lasers described thus far are characterized by a *homostructure* junction; that is, both sides of the junction are composed of the same semiconductor material. The heterostructure laser employs a substrate of GaAs, to which a layer of aluminum gallium arsenide [(AlGa)As] is epitaxially grown. Since GaAs and (AlGa)As have nearly identical crystal structures, it is possible to grow a very uniform heterojunction of the two materials.

Typical heterostructure lasers have a threshold of only about 8000 A/cm^2, about 20% of the lowest value for good epitaxial homostructure devices. This dramatic reduction in required current density results from the index of refraction difference between GaAs and (AlGa)As. Since the refractive index of the latter is higher, light generated along the pn junction is prevented from escaping into the light-absorbing p region by being reflected back into the active region. Light can still be lost in the n region, but the improvement in optical confinement is so significant that a substantial lowering of the required current threshold results.

The slight bandgap difference between (AlGa)As and GaAs gives rise to an electron confinement which also helps laser action. The slight energy difference establishes a potential barrier which tends to confine stimulated electrons to a thin region along the junction. The combination of optical and electron confinement results in a highly efficient, low-threshold laser structure.

An obvious improvement of the heterostructure laser would be the addition of a second (AlGa)As layer on the n-side of the junction. A diode with one (AlGa)As-GaAs junction is called a *single-heterostructure* (SH) laser, while a (AlGa)As-GaAs-(AlGa)As structure is called a *double-heterostructure* (DH) device.

The DH laser was first constructed in the Soviet Union. But the most impressive accomplishment with this new configuration occurred in 1970, when a Bell Laboratory team (headed by I. Hayashi) fabricated DH lasers with a threshold of only 1000 A/cm^2. Prior to this historic development, injection lasers could not be operated continuously at room temperature due to the heat buildup caused by the required current densities. The new DH lasers permitted continuous operation at room temperature and above for the first time.

RCA was working with DH lasers at the time of the Bell Laboratory achievement and soon devised continuously operating lasers also. RCA made a further contribution by devising a modified DH laser employing a thick optical emission region. Conventional SH and DH injection lasers have a very thin active region. This results in a very high power density along the emitting region of the junction, and device destruction occurs if power-output levels are too high. The destruction results from melting or small pieces of the reflecting crystal surface are literally blown away by the high-power density. Similar destruction occurs in high-power glass, ruby, and other solid lasers.

The new RCA diode is called a Large Optical Cavity (LOC) laser. Since the emitting region is thicker than that of conventional lasers, higher power outputs than DH lasers are available without device degradation or destruction. A commercial LOC laser is shown in Fig. 7-3.

ELECTRO-OPTICAL PROPERTIES

When a small current is applied to an injection laser, the device acts very much like a light emitting diode. In fact, any semiconductor which can be fabricated into an injection laser can be made into a high-quality LED.

As current through the laser is increased, the spectral output begins to narrow. As the population of stimulated electrons becomes inverted, optical gain exceeds optical loss, and the diode begins to lase. The spectrum has now narrowed considerably and may be only a few nanometers wide. Also, the light which was formerly emitted from all portions of the chip not blocked by electrodes emerges in a distinct beam from the laser end mirror.

As the current is increased above threshold, more modes begin to propagate in the laser, and the output spectrum widens to perhaps 30 nm. The power output increases linearly with applied current until the laser chip becomes overheated. Like conventional LEDs, the power output then begins to fall.

The optical power output of an injection laser as a function of injection current is shown in Fig. 7-4. Note that the output below the lasing threshold is linear until the threshold is reached. The output then turns sharply upward and again resumes a linear output but with a much steeper power curve.

Fig. 7-3. High-efficiency LOC injection laser diode.

Typical injection lasers emit from one-half watt to 50 watts in pulsed operation at room temperature. Values as high as 100 watts have been reported for wide devices mounted on large copper heat sinks. Homostructure and SH lasers emit far more power than DH devices. The emission region of the latter device is so small that power densities sufficiently high to cause mechanical damage are easily obtained.

An important characteristic of the injection laser is the containment of its radiation within a distinct beam. Most solid and gas lasers have a very narrow beam divergence, but since the light from an injection laser emerges from what is essentially a small slit (the junction), it is diffracted outward into a beam with a typical divergence of 25°. The DH laser have a thin active region and their divergence is about 40°.

The beam from an injection laser can easily be collimated with a small, inexpensive lens. The lens should have a small f/number to ensure that most, or all, of the laser light is collected and projected. The simple laser systems described later in this chapter can be adjusted to produce a spot of invisible infrared only 5 mm by 120 mm at a distance of 15.25 meters. This corresponds to a divergence of 0.33 mr by 7.87 mr (0.02° by 0.45°). This divergence can be easily increased by moving the lens closer to the laser.

Another important characteristic of injection lasers is their narrow spectral width. While typical LEDs have a spectral output width of some 30 or 40 nm, diode lasers have a spectral width of only 3 or 4 nm. The width is even narrower just above threshold in most lasers and well above threshold in specially fabricated devices.

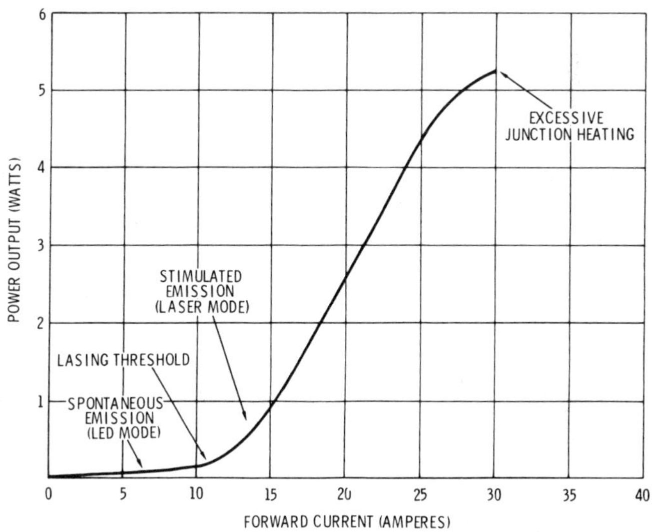

Fig. 7-4. Injection laser current versus power output.

Finally, a discussion of the visible nature of injection laser radiation is appropriate. The upper limit of human vision is typically specified as about 720 nm, but this author and others have reported that GaAs lasers operating at room temperature and emitting at 905 nm are visible as a red point of light. When cooled with liquid nitrogen, a GaAs SH laser, emitting at about 850 nm, becomes a bright cherry red and the beam is clearly visible in a well-lighted room. High-power arrays of GaAs lasers operated at room temperature sometimes appear violet in color. The author has studied this phenomenon and ascribes it to a mixture of the red at 905 nm and a second harmonic in the blue at about 452 nm. One study of a diode, emitting 20 watts of infrared, detected 10 microwatts of blue. This much light is definitely visible and provides a likely explanation for the purple or violet appearance of light from some injection lasers.

LASERS VERSUS LEDS

Certain applications for LEDs, such as indicators and numeric displays, cannot be performed by lasers. The reverse is also true. For example, LEDs cannot perform as efficiently in long-range optical communication links as diode lasers.

There are, however, certain applications where both a laser and LED can be used, and where a careful consideration of the relative advantages and disadvantages of each is essential. Voice communicators, intrusion alarms, and miniature ranging and detection systems are typical examples. Some of the relative merits of each device have been discussed in this book and most can be summed up by comparing a typical laser and typical LED. Consider, for example, an injection laser capable of delivering 4.5 watts when driven with a 10-ampere pulse and an efficient, inexpensive LED capable of delivering 6 milliwatts at 100 milliamperes bias. Representative commercial devices with these specifications are, respectively, the RCA SG2002 laser and the General Electric SSL-55C LED.

When driven with a 10-ampere pulse, the LED will deliver about 240 milliwatts, only about 5.3% of the 4.5 watts emitted by the laser for the same peak forward current. Equally significant, the power-conversion efficiency of the LED has dropped from about 5% at 100 milliamperes forward bias to only 1.7% at 10 amperes. The power conversion efficiency of the laser is an impressive 5.6%.

In addition to high peak power and conversion efficiency, the infrared emerging from an injection laser is far easier to collimate than that from an LED. This is because the injection laser emission is directional and emerges from a region on the laser chip only about 0.002 mm by 0.076 mm. Radiation emerging from a source this small can easily be collimated into nearly parallel light with a very small f/1 lens.

Depending on the mechanical structure of the LED, the lens which produces a highly collimated beam for an injection laser gives a practical minimum divergence of about three degrees when used with an LED. Divergence can be reduced further, but only at the expense of optical power contained within the projected beam. Worse, only 20% or less radiation from the LED can be collected by a simple lens system. Parabolic reflectors are needed for higher collection efficiency.

Thus far, the laser appears to be a superior radiation source. However, its driving requirements lessen its overall advantages. In order to fulfill the requirement for sustained laser oscillations, an inversion in the population of stimulated electrons, the injection laser must be pulsed with current levels as high as 100,000 amperes/cm^2. Recently, as noted earlier in this chapter, low-threshold, double-heterostructure devices have been produced which have reduced required driving current to only a few thousand amperes per cm^2, but these devices have shorter operating lifetimes. Practical operational current densities for good-quality, single-heterostructure injection lasers range from 40,000 to 50,000 A/cm^2. In short, to achieve the population inversion prerequisite to laser action, very high current densities must be present in the semiconductor junction.

To reduce the current to practical levels of from about ten to one-hundred amperes, injection lasers are made quite small. The RCA SG2002 described above, for example, has a junction area of only 0.076 mm by 0.305 mm. This reduces the required current to about 10 amperes, but the tiny volume of the laser cannot sink this current, if applied continuously, without being destroyed by thermal action. For this reason, conventional nondouble-heterostructure injection lasers must be operated at pulse widths of typically no more than 200 nanoseconds. To avoid a gradual pulse-to-pulse thermal accumulation, duty cycle (total "on" time) must not exceed 0.1%.

These driving requirements mean great care must be exercised in the design of the injection laser driving circuitry. For most practical considerations, two options are available to obtain the necessary high current and fast pulse width. One is to switch a high voltage through a controlled semiconductor switch such as a silicon controlled rectifier (SCR). The other is to switch a voltage through an avalanche semiconductor switch such as a selected transistor or four-layer diode. In both cases, the voltage is stored in a small capacitor until it is discharged through the semiconductor switch and laser.

The SCR technique has the advantage of simple design and operation, but typical SCRs present a relatively high impedance to the very fast pulse required to operate a laser. For this reason, several hundred volts are necessary to obtain the 20 to 30 amperes required to drive conventional single-heterostructure lasers.

Avalanche transistor circuits can be designed which produce similar current levels for significantly smaller voltage levels. This is because avalanche transistors possess a much lower impedance at fast pulse widths. In a comparison of practical SCR and avalanche transistor driving circuits, conducted by the author, typical SCRs possessed an impedance of nearly 5 ohms for a 75-nanosecond pulse. Typical avalanche transistor impedance was only one-third this value.

Obviously, diode laser pulsers require a careful study of trade-offs before selection of a final circuit. Even with great care in circuit design, a laser circuit will consume more power than a similar LED circuit.

Injection lasers present other special problems. For example, the fast pulse width of an injection laser imposes special design requirements on a receiver. For most practical considerations, only reverse-biased pin or avalanche photodiodes, coupled into a wideband amplifier, make for a practical receiver combination. Also, conventional lasers have a limited lifetime, and their degradation is continual and irreversible. Good-quality, single-heterostructure lasers have demonstrated improved lifetimes. A study by RCA has shown lasers, operated at 1 kHz, possess 80% of their original output after 1000 hours operation.

Another disadvantage of injection lasers is temperature sensitivity. Typical lasers exhibit significantly lowered current-threshold values as temperature drops. A system designed to operate a laser at, or close, to its maximum current level (to achieve maximum power output) will fail if the temperature of the laser drops, causing its maximum permissable driving current level to fall below that delivered by the driving circuit. Depending on the temperature change, the laser may be degraded or even destroyed. In a typical system where the laser is operated at its maximum current rating, a drop in temperature of only 10 °F causes laser failure.

Temperature regulation can be employed to protect the laser, but a simpler procedure is to limit peak current to a safe value for a specified temperature range. The laser will simply not be operated when ambient temperatures exceed the allowable range.

Other potential difficulties with injection lasers include electromagnetic pulse (EMP) effects and safety. The EMP phenomenon results from the very fast, high-current discharges required to drive a laser. EMP can cause substantial interference in nearby electromagnetic receivers such as radios and sensitive, aircraft navigation equipment. Fortunately, the effects are generally minor. In a test conducted by the author, an injection laser system produced no discernable interference in an a-m radio receiver, so long as the two were separated by at least 75 cm.

Safety is a less precise area. The high voltage (e.g., several hundred volts) which may be necessary to operate some laser systems poses a potential hazard, but only if the protective housing of the device is opened. Of more concern is the potential ocular hazard to persons within the field of view of the laser. This topic is discussed later in this chapter.

Summing up, the laser is superior to the LED in power output and ease of beam collimation, but its driving requirements can pose substantial difficulties. Though LEDs offer less peak power and optical collection efficiency, they are nevertheless useful in low-cost, compact equipment requiring moderate power levels. Each case, of course, must be considered on its own merits.

INJECTION LASER OERATION

Lasers of the DH-type will operate continuously at room temperature and supply up to 100 milliwatts for an input current of one ampere. The power supply for a DH laser can consist of a low-voltage, high current capacity battery and a current-limiting resistor. A current meter in series with the laser will permit the laser current to be monitored. As when operating a high-power infrared LED, it is advisable to monitor the output of a DH laser as current is increased. When the optical output begins to taper as current is increased, the current should be reduced by 10 or 20%. Commercial DH lasers capable of room temperature continuous operation are not yet available. Manufacturers will supply appropriate driving information when they do become available.

Conventionally heat-sinked LOC lasers cannot be operated continuously at room temperature, though large copper heat sinks on both sides of the laser chip can be used to obtain cw operation. Homostructure and SH lasers also cannot be operated continuously at room temperature. All these lasers require special current-pulsing drive circuits for safe operation. Current-pulsing circuits add to the complexity of an injection laser system, but a beneficial result is very high peak-output powers.

AVALANCHE TRANSISTOR PULSE GENERATOR

A very simple laser pulser can be made with a switching transistor, operated in its avalanche mode. By simply increasing the voltage across the collector-emitter leads, until breakdown occurs, very fast pulses can be generated. Not all switching transistors will operate in the avalanche mode, but many will.

A simple avalanche transistor, laser pulser is shown in Fig. 7-5. In operation, C1 charges through R1 and R2, until the breakdown voltage of Q1 is reached. Base bias to Q1 is provided by R3. When Q1 avalanches, the voltage stored in C1 discharges through Q1, the laser diode, and R4. So long as the peak current rating of the laser diode is not exceeded, current monitor R4 may be omitted from the circuit.

This circuit is a relaxation oscillator and delivers a repetitious series of laser pulses automatically. As C1 discharges, Q1 turns off and C1 can again charge through R1. When C1 reaches the avalanche voltage of Q1, Q1 switches on and the process repeats.

This laser pulse generator can be assembled on a small circuit board using either etched circuitry or point-to-point wiring. Prevent inductance, which will increase pulse width and cause current undershoot, by keeping the leads in the discharge circuit as short as possible (see Fig. 7-5).

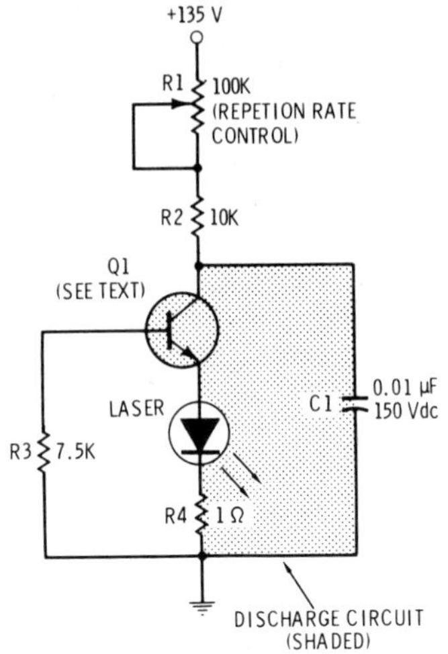

Fig. 7-5. Avalanche transistor laser pulse generator.

Many different types of transistors will operate in the circuit; good choices include the 2N918, 2N222, 2N3643, and other npn silicon switching transistors. Before a laser is connected to the circuit, the peak current delivered by the pulser must be carefully measured to prevent possible damage to the laser. If a fast oscilloscope (15 MHz) is available, this can be done by connecting a one-ohm resistor in place of the laser as a dummy load while monitoring the current through R4. Different transistors will require various power-supply voltages and will deliver different currents. For example, 2N2222

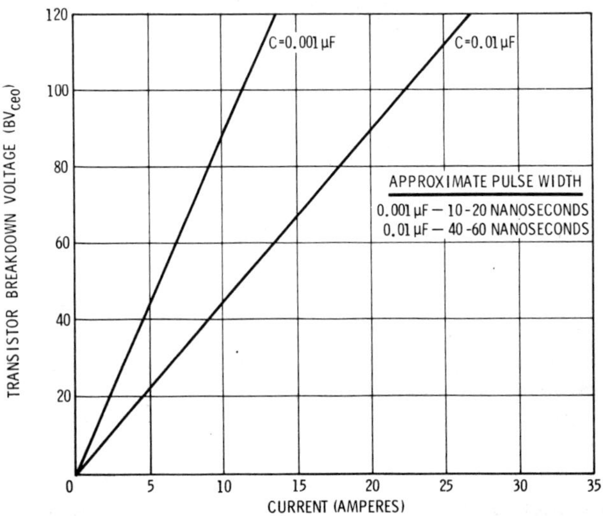

Fig. 7-6. Peak current versus avalanche breakdown voltage.

transistors typically deliver 15 or more amperes in the circuit, while 2N3643 units deliver half this value.

When a transistor which gives the required laser current has been selected, connect the laser into the circuit and once again measure the current through R4. The current will be slightly different than with the dummy load.

If a fast oscilloscope is not available, measure the peak current indirectly. Connect a voltage-calibrated, low-speed oscilloscope across C1 with a dummy one-ohm load resistor substituted for the laser. The scope should display a sawtooth wave representative of the charge-discharge cycle of Q1. The voltage amplitude of the wave corresponds to the breakdown voltage of Q1. Fig. 7-6 is a graph of approximate peak current as a function of the breakdown voltage of Q1 for two values of C1. This graph cannot be relied upon for extreme accuracy, due to variations in the tolerance of circuit components, particularly C1, but it does provide a general indication of peak current.

SCR PULSE GENERATOR

The simple avalanche transistor, laser pulser is ideal for portable applications, but is limited in usefulness since peak current cannot be varied without substituting transistors and rechecking laser current. Much better results can be obtained with an SCR pulse generator. Since the SCR can be externally triggered, laser current can be varied from very low to very high levels.

A circuit for a SCR complete-injection laser pulser is shown in Fig. 7-7. This circuit was designed by the author for *Radio-Electronics* magazine, and it originally appeared in the June 1972 issue. The completed laser is shown in Fig. 7-8.

Operation of the circuit is straightforward. The circuit uses 1:1 isolation transformer T1 with a 6.3-volt secondary winding. A voltage doubler, which converts the 125 Vac from T1 to 380 Vdc, is formed by C1, C2, D1, and D2. This voltage is used to drive a charging circuit composed of Q2 and a resistor network. The charging circuit charges C7 to any desired voltage up to 380 volts prior to being discharged through the SCR and the laser diode.

The SCR is triggered by means of a simple unijunction oscillator circuit formed around Q1. Power for the oscillator is supplied by a second voltage doubler composed of C3, C4, D3, and D4. This doubler converts the 6.3 Vac from the secondary of T1 to 19 Vdc.

When the UJT oscillator triggers the gate of the SCR, C7 discharges in a rapid, high-current pulse through the SCR, R7, and the laser. Current undershoot through the laser is prevented by bypassing any negative voltage through D5. Current monitoring is accomplished with R7.

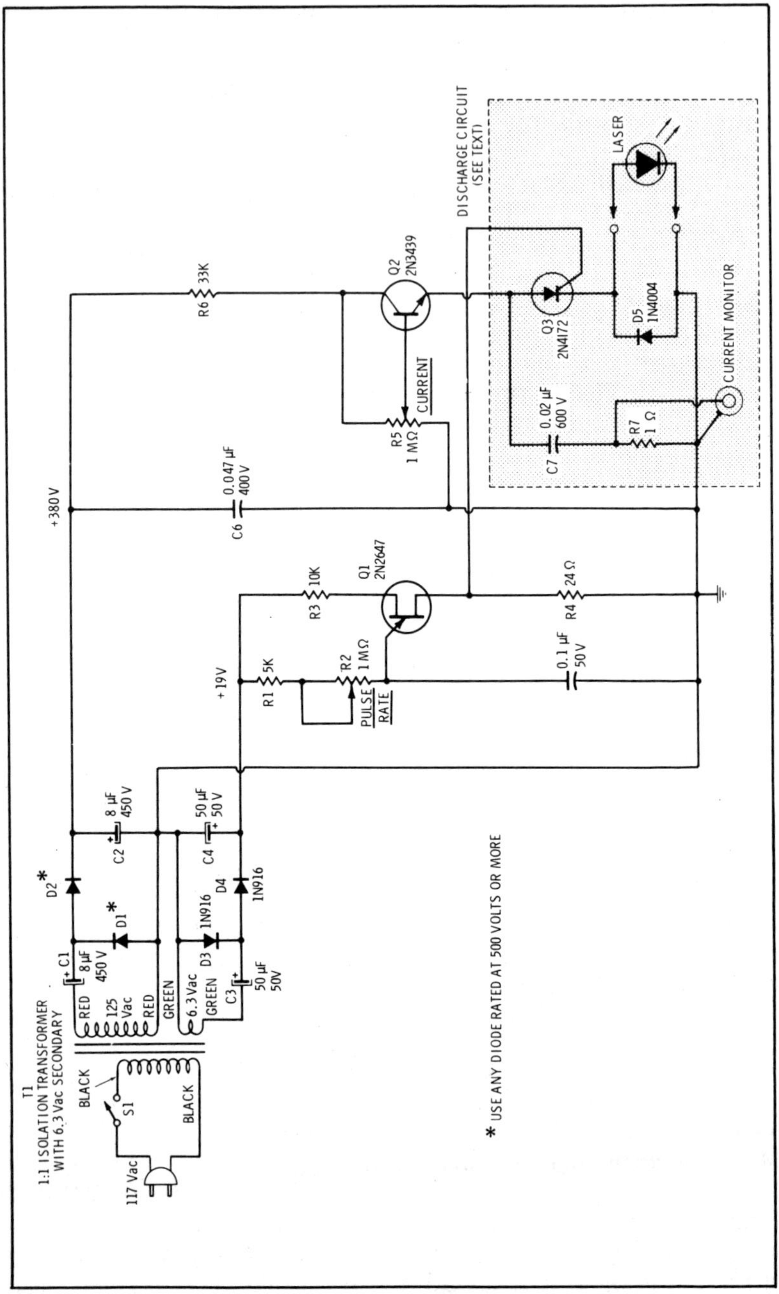

Fig. 7-7. SCR laser pulse generator.

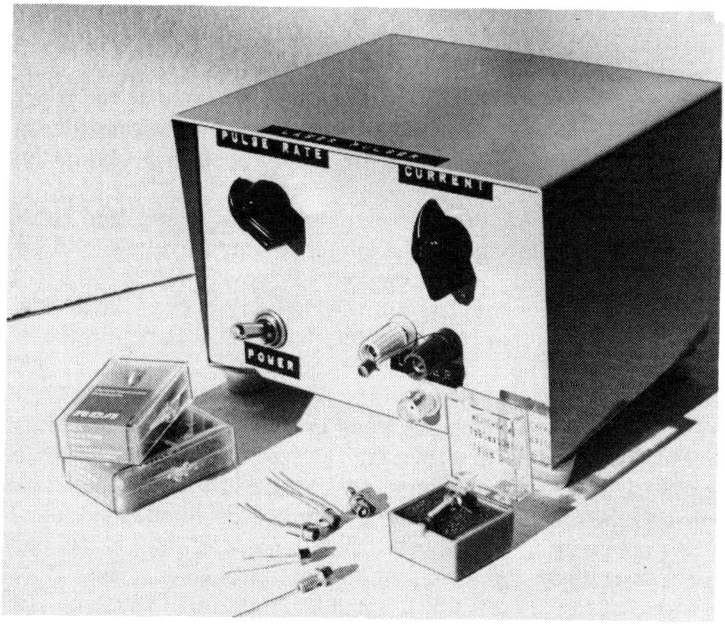

Fig. 7-8. Variable current and pulse rate, injection laser pulse generator.

The circuit can be easily duplicated on a perforated board. Mount the board in a cabinet to facilitate operation and install current control R5 and repetition rate control R2 in holes drilled in the housing. The discharge circuit must be wired with short leads to prevent distortion

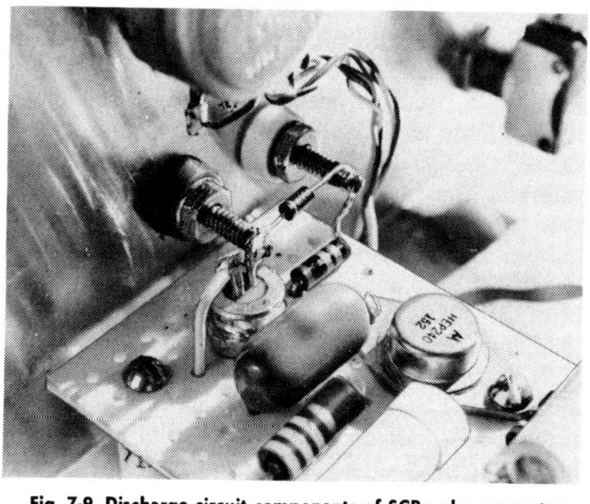

Fig. 7-9. Discharge circuit components of SCR pulse generator. Note point-to-point wiring to reduce inductance.

of the laser current pulse. Fig. 7-9 is a photograph showing the layout of the discharge circuit in the prototype pulser. The two, threaded terminals form the laser connection points. Note how D5 is soldered directly across both terminals and how the cathode of the SCR is soldered directly to one terminal. The point-to-point wiring illustrated in Fig. 7-9 keeps the discharge current path about five centimeters in length.

All components for the pulser are readily available, and substitutions can be used throughout the circuit so long as voltage and power ratings are observed. Four components, however, are critical and should *not* be changed unless a 15-MHz oscilloscope is available for independent calibration of the circuit. The critical parts are the SCR, C7, R7, and D5. The SCR is a specially selected unit, and only higher voltage units in the same series may be substituted. Other SCRs may have a slow rise time which can result in stretching of the current pulse and excessive heating of the laser. Discharge capacitor C7 has been selected to provide the maximum possible pulse width to the laser and it *must not* be increased in value. The prototype circuit was calibrated via R7; removing it, even though it has a value of only one ohm, will cause substantially more current to flow through the laser. Finally, D5 is one of few diodes which adequately absorbs reverse current in a laser pulser.

When the circuit is complete, test it by connecting a one-ohm resistor across the laser terminals and place a radio near the unit. A buzzing sound, originating from the discharge circuit, should be heard from the radio when the pulser is turned on. Discharge capacitor C7 usually emits a buzzing sound as well, and this sound will verify operation of the circuit.

If the pulser does not operate, check the power supply circuitry for correct operating voltages. Also make sure that the UJT oscillator is operating. When the circuit is operating properly, reduce the current to zero, install a laser in the laser terminals (being sure to observe correct polarity), and calibrate the instrument.

Three calibration techniques are available. The best is to connect a 15-MHz, or faster, oscilloscope to the current monitor and directly measure the laser current. Since the monitor is one ohm, one volt on the scope will equal one ampere of current through the laser. As the current control is adjusted upward, mark the front panel of the housing at one-ampere intervals.

CAUTION: Do not increase the current above the maximum value allowed for the diode laser connected to the pulser or the laser may be damaged or destroyed. A higher-current laser may be required to calibrate the pulser at its maximum output current.

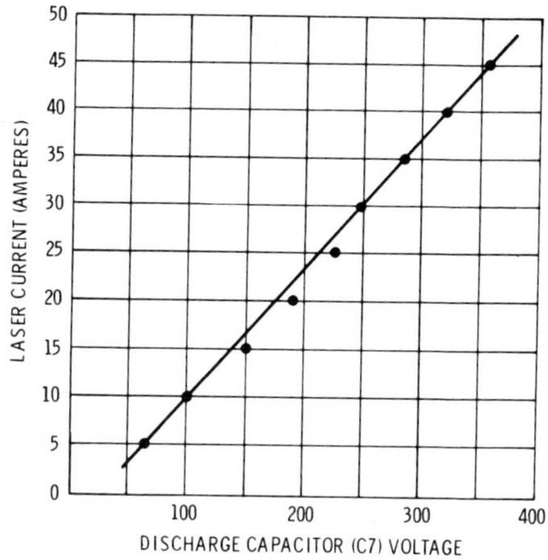

Fig. 7-10. SCR laser pulse generator calibration graph.

If a fast oscilloscope is not available, a voltage-calibrated scope of any speed can be used. Connect the scope across C7, with the pulser turned off, and then activate the pulser. The amplitude of the charge-discharge wave is related to laser current as shown by the graph in Fig. 7-10. The graph is valid only if no parts substitutions have been made in the discharge circuit.

A third way to calibrate the pulser does not require an oscilloscope. The pulser is turned off and a vom is connected from the rotor of R5 to ground. The voltage is related to the laser current, as shown by the graph in Fig. 7-11. Again, the graph is valid only if no parts substitutions have been made in the discharge circuit.

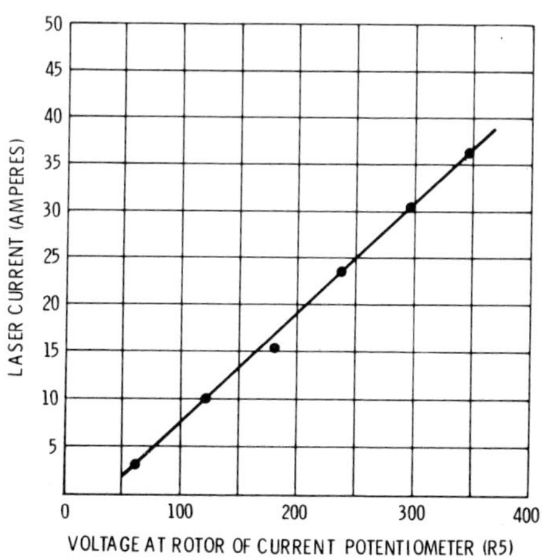

Fig. 7-11. SCR laser pulse generator calibration graph.

OTHER PULSE GENERATORS

Laser pulse generators can be designed around a host of switching devices, ranging from mercury-wetted relays to miniature thyatrons. Four-layer diodes and conventional transistors can also be used. In all cases care must be exercised to prevent exceeding the duty-cycle and current ratings of the laser. For a comprehensive discussion of laser pulsers, including circuit diagrams, see *Semiconductor Diode Lasers,* by R.W. Campbell and F.M. Mims (Howard W. Sams & Co., Inc., 1972).

LASER SAFETY

The electrical hazards of laser power-supply equipment are well known. However, because of the difficulty, variability, and expense of biological experiments to measure the precise quantity of laser radiation required to produce a retinal lesion, ocular damage levels are not precisely defined.

As a general guideline, the Air Force School of Aerospace Medicine lists 0.75 microjoule as the maximum permissible energy to enter the eye from a ruby laser (694.3 nanometers) with a 100-nanosecond pulse width. For a neodymium-doped laser (1060 nanometers), the permissible energy level is 45 microjoules. Injection-laser safety levels fall somewhere between these numbers.

The large difference in safe exposure limits between ruby and neodymium results from absorption of infrared in the vitreous humor of the eye and imperfect focusing of infrared wavelengths. Furthermore, since the beam from an injection laser broadens significantly, it would be difficult to receive even the 0.75 microjoule level without placing the laser directly adjacent to the eye.

In the case of a well-collimated GaAs laser beam, more energy can enter the eye. Therefore, to avoid any possibility of ocular injury, always follow these simple safety rules:

1. Treat the injection laser as any other bright light source and do not look directly into its beam.
2. Do not allow collimated light from an injection laser to strike mirrors or other shiny surfaces.
3. Warn assistants and onlookers when preparing to conduct a test or experiment with an injection laser.

For additional information on the subject of laser safety, obtain a copy of *Laser Fundamentals and Experiments,* written by the Bureau of Radiological Health (Department of Health, Education, and Welfare). It is available for $1.25 from the Superintendent of Documents, U.S. Government Printing Office, Washington, DC 20402.

Index

S

T

V